Socialist States and the Environment

'A fascinating account. For too long we have tended to demonise socialist states, this book shows that to overcome the climate crisis, there are positive lessons to be learnt, from Lenin's promotion of conservation to Cuba's achievements in promoting ecological policymaking.'

—Derek Wall, former International Coordinator of the Green Party of England and Wales, and Lecturer at Goldsmiths, University of London

'Many people have realised that ecological sustainability cannot be achieved under capitalism. But how about (eco-) socialism? For everyone who is interested in a sustainable future and a new society without oppression, I strongly recommend this book.'

—Minqi Li, Professor of Economics, University of Utah and author of *China and the Twenty-First-Century Crisis*

'In our current moment of a near total co-optation of environmentalism where billionaires and military forces are looked to for solutions to the problems they create, Salvatore Engel-Di Mauro provides a serious, data-driven, and sober look at what socialist states have been able to do for the environment.'

—Justin Podur, Associate Professor at the Faculty of Environmental and Urban Change, York University

'Sharply erudite … takes us on a brilliant guided tour of the environmental programs of socialist states and a variety of community-led initiatives. Among them, Thomas Sankara's Burkina Faso, Cuba and its agroecology, the PRC up to 1978, the USSR and many of the Eastern European countries up to 1990, various African people's republics through 1992, and, despite their largely privatised economies, the Bolivias, Venezuelas and Vietnams of today.'

—Rob Wallace, author of *Big Farms Make Big Flu*

Socialist States and the Environment

Lessons for Ecosocialist Futures

Salvatore Engel-Di Mauro

First published 2021 by Pluto Press
345 Archway Road, London N6 5AA

www.plutobooks.com

British Library Cataloguing in Publication Data
A catalogue record for this book is available from the British Library

ISBN   978 0 7453 4041 8   Paperback
ISBN   978 0 7453 4040 1   Hardback
ISBN   978 1 78680 789 2   PDF
ISBN   978 1 78680 790 8   EPUB
ISBN   978 1 78680 791 5   Kindle

This book is printed on paper suitable for recycling and made from fully managed
and sustained forest sources. Logging, pulping and manufacturing processes are
expected to conform to the environmental standards of the country of origin.

Typeset by Stanford DTP Services, Northampton, England

Simultaneously printed in the United Kingdom and United States of America

Contents

Figures

Figures

Tables

Preface

Down from Mars mountain, in the Biellese Alps of the north-western Italian peninsula, there runs an intermittent stream, the Elv (Elvo in Italian). It is swollen by the Olobbia, Oremo, Viona and Ingagna creeks along its way to become a tributary of the river Cervo, among the many feeders of the Po. Long ago, through multiple phases of glacial erosion in the Early Pleistocene (one to two million years ago), these and other now bygone streams carried and dumped auriferous pebbles, gravel and sand along their courses to form a bulging, oblong strip of land, a couple of river terraces layered on top of each other.

This is the Bessa highland, a natural preserve since 1985. With what turns out to be an exiguous amount of that shimmering metal, the Bessa is part of a much larger gold-bearing area dotted with tiny ancient mines. It rises between 300 and 450 metres above sea level, an infinitesimal altitude next to the nearby towering Alps. More than a couple of millennia ago, this was Salassi domain before a Roman contingent led by Consul Appius Claudius Pulcher had the better of them (143–140 BCE) and secured the gold-bearing land for the Roman Empire. Barely a century later the gold-bearing deposits of the region were deemed unworthy of more mining efforts.

To this day, assortments of mainly hobbyists and fortune seekers converge on the Bessa, sifting through sediment, looking for miracles, though of different sorts. It is said that the Victimuli (presumably, in part, the descendants of the Salassi) melted much of the gold to form the statue of a horse, which they interred in the higher portion of the Bessa to conceal it from the Romans. Centuries later, fairies allegedly visited this stretch of once gold-bearing land and promised to share their talent for finding gold. The locals could not have been happier with such a revelation, but during a feast they made the unpardonable error of ridiculing the fairies for their anserine feet. Deeply offended, the fairies left, never to return.

The magical Mars mountain, the quick and lively interlocking montane streams, the glacial refashioning of land yielding concentrations of gold, ancient land struggles, the skilled fairies, the modern breakthrough-seekers and the best means to get that elusive gold. This is the long, rocky, slippery, shining, trap-riddled path of recovering and rebuilding socialism, including anarchism and communism, the precious substance that is within our grasp and made elusive by both malicious and well-meaning forces. Recalling and rediscovering bits of history and legend from one's land of birth can be a salutary method to recover what seems to be lost, but is in reality far from it.

Shorn of petty parochialism and nostalgia, this is itself a challenge of overcoming. It is a process of mixing the grounding in a place – familiar yet, for someone like me, alienated – with the ethereal flight necessary to reach beyond the concreteness of the present and the restraints of the past, including the place of origin, which allows, if all hinges well, for a return to the same place, even if not physically, with renewed vision and reinvigorated determination.

The endeavour implies the opposite of seeking fortunes or collectible relics in a debris of prematurely discarded history. In other words, the sifting process is one of looking for clues for what to do in what we know well or thought we did, and in the revisiting process find out we did not know as well as we thought or appreciate as well as we ought. Whether such renewal makes for any improvement in perspective and political struggle is another matter. The reader must decide that, ultimately.

At least, for what it is worth, it is how I would like to conceive of my political journey thus far and what I want to share with you, though henceforth not in such an allegorical way. One could start such a journey from a much benighted present and work backwards. But working forwards from the past is an essential complement to comprehend the current state of affairs. This is where the importance of various historical socialist experiences becomes evident, especially as they are treated like dirt from both right and left, even as gold cannot exist as such without the dirt that bears it.

It could be that clue seared in the back of my head from the permanent strained look and buried frustrations on the faces of some gimnázium colleagues I saw in the early 1990s while living on teachers' wages on Balaton Lake, contrasted by the verdant countryside and plentiful woods I enjoyed, along with their stores of delightful mushrooms and tea herbs.

It must be that heated word slinging over the merits of life under socialism between two villagers at a dusty and bustling village bus stop or that nostalgic twinkle I could discern in some farmers' eyes when talking about the Kádár government, while I was living in south-west Hungary in the late 1990s, and the frenetic pace and copious use of whatever agrochemicals one could find to meet the new watermelon production targets set by some unknown bosses far away in Sweden or Greece.

It must be the expression of political indifference among the Roma I met and the grinding poverty at the margins of rural villages, with their fresh abusive and smelly dumping grounds, and the ride given to a day worker just sacked and stranded tens of kilometres from home and without the money to get there via an ever more expensive and rarefied public transport system, while the financial heist of 2008 raged on, mercilessly hacking to bits the life savings of millions.

It must be that 2010 toxic red mud disaster at Ajka, in Hungary, killing at least seven villagers, rekindling home-town childhood memories of toxic spills contaminating groundwater and sending folks in droves to the shops for bottled water and of anxiety-sparking events not too far away, like the Seveso accident and dioxins terror. Only much later did I learn of the shocking 1963 Vajont Dam disaster, when a massive landslide resulted in nearly 2,000 people drowned or suffocated to death by a sudden rush of muddy water.

But it was the profoundness of the words of a couple of Baré and Jivi (Guahibo) activists in Yaracuy (Venezuela) that added the final displacing weight needed to start unravelling like a load of bog rolls all my prior, sedimented (pre)conceptions of 'Soviet' dictatorship, of authoritarian socialism, of leftist statism. They, too, were critical of Chavismo, and yet they were convinced Chavistas. They recognised and lived the contradictions, but also the necessity to continue along the path taken, seeing also the tangible betterment of the lives of their communities that would not have been possible without the Chávez government. They saw the huge strides being made as well as the setbacks and unsettlingly uncertain future. The former must

always be underlined, and the latter must be faced with the utmost
clarity and commensurate vehemence, rather than used to dismiss
all the gains made and to foreclose any like path in the future. It is all
part of the messy, at times regrettably violent struggle, where taking
one's eyes off the ultimate objectives means self-erosion, capitulation
and reabsorption into the even more detestable, into an even worse
set of conditions. This is what I took from the conversations I was
honoured to have with them.

And the Baré and Jivi, as with so many Indigenous communities
and peoples struggling for decolonisation, know very well what is at
stake, which, unlike for a community like mine, is survival. Up to that
point, a full four years after the untimely passing of Hugo Chávez,
I had followed the wagon of aloofness, scepticism, critique and
refusal of anything combining the state and socialism or communist
struggle, a catastrophic oxymoron to me in a previous life and to that
of so many others still. Chavismo was just another expression of the
wrong-headed kind of socialism that inexorably fails, gets distorted
or corrupted, and ends up in some authoritarian Hades. And destroys
the environment as well, just look at that greenhouse gas-belching oil
economy.

I was not just wrong, I was thoroughly misguided and misdirected,
unable to see through the privileges in which I was raised in liberal
democracies and the projection of frustration on revolutionaries in
the rest of the world of the immense historical defeats and, in some
cases, resignation that the left in the imperialist centres had suffered
and internalised. I was and to some extent still am among them, the
projectors, imbibing that mellifluous labour-aristocracy drink of
self-rot while trying to make myself vomit it out for good.

This book is not an exercise in finding the praiseworthy in 'socialist'
states, but about reconsidering their environmental impacts according
to wider global and ecological contexts, to try to understand what
challenges can be expected in the struggle for socialism and what can
be done differently for a socialist future. It is a response to the still
predominant and facile capitalist and environmentalist rhetoric that
would have us equate socialism, and even more so communism, with
'ecocide' and 'totalitarianism'.

In some respects this work is also an attempt to understand what made state-socialist societies into poor or spent alternatives. In general, it should be household knowledge that the terrible outcomes that came out of state-socialist countries are as justifiable as those that have always typified free-market 'democracies'. All those who favour capitalism or promote capitalist 'democracy' must be tasked with explaining how they can support social systems predicated on historical and current horrors. I suspect this will not happen until the forces sustaining capitalist regimes are defeated and, more desirably, a classless egalitarian order is established.

What is being recommended here, instead of dismissing socialist states as not socialist or as proof that socialism is fatally flawed, is to grasp the reasons for the trajectories socialist states took and to understand the contexts through which those trajectories happened. Without doing so it is not possible to oppose effectively the massive and overwhelming disinformation on socialism (including within parts of the self-described left) and to draw up workable political alternatives and strategies to overcome capitalism (the most ecologically destructive system in the history of humanity), prevent further planetary destruction and embark on building an ecologically sensible society.

The task is not to formulate any blueprints for the future. That, even if legitimate, would require first the end of capitalism in at least most of the world so as to establish globally coordinated popular assemblies deciding on and writing up place-sensitive guidelines. Obviously this is rather far from the currently existing political conditions. Alternatively one can, indeed must, learn from the past regarding what kinds of situations and actions can lead to what kinds of outcomes and legacies so as to do the utmost to prevent or at least minimise the carnage and environmental devastation associated with many of the socialist revolutions that temporarily prevailed and then foundered or were made to founder.

The obverse of this task, not undertaken in this volume, is to demonstrate and underline continuously, insistently and repetitively the horrors of liberal 'democracies' and capitalism generally, since liberal democrats and capitalist encomiasts will not take ownership of or will not feel responsible for the disaster that is their preferred kind of society and since exactly such demands continue to be made of socialists of any persuasion. In this adverse situation, any concessions to such ideologues must be resisted without falling into any idealisation or rationalisation of past and current socialisms. To accomplish this, the realities of socialism need to be confronted, not ignored or dismissed as an alien other.

Arguing with anti-socialists and especially anti-communists about getting communism right is not too different from arguing with anti-anarchists about desisting from defining anarchism as chaos. It is likely a fruitless effort because political antagonists of egalitarianist movements are usually more interested in maintaining power or in maintaining their privileges (whether explicitly articulated or not), than any logical and evidence-based debate. In such a mindset the point is to debate only insofar as it furthers the conquest of political power, not about convincing or educating anyone or sharing a different perspective on the same problem.

But I would wager that, as in the case of the biophysical and technical sciences (also known as science, technology, engineering, and maths), most simply do not know much, if anything, about socialism, communism or anarchism. Educating, agitating, debating and convincing should be activities of primary importance in that regard. The battle of ideas, if one wishes to make analogies to war, is really waged against self-aware antagonists and not with most people, who might even see arguments of this sort as senseless.

This book is then not an attempt to convince those who already (think they) know, but to share a viewpoint with those who wonder about socialism or who are open to differing interpretations of socialism and its histories. It is certainly not some innocent sharing process on my part. As anyone, I have a set of principles and a related (or consequent) political proclivity, which is here identified as an anarchist-communist and (eco)feminist variant of ecosocialism, which may not always be evident in the ways I express myself in what follows.

Acknowledgements

Just conceiving of this book took a couple of years. Its precursor is a 2016 editorial appearing in *Capitalism Nature Socialism*, "The Enduring Relevance of State-Socialism". The further development of that editorial into a larger project would not have been possible without the solicitation and editing care of Associate Commissioning Editor Jakob Horstmann. Along the way, the work benefited greatly from the critical eyes of Deborah Engel-Di Mauro, which was especially important to what became Chapter 2 of this book, and of Marco Armiero, Leigh Brownhill, Maarten DeKadt, Danny Faber, Mazen Labban, George Martin, Laura Pulido, Jose Tapia and Judith Watson, who all contributed in different ways to the making of that 2016 editorial and what came of it afterwards. Robert Webb's and David Castle's supervision and Dan Harding's copy-editing at Pluto Press have greatly facilitated this work's completion. I cannot thank all of them enough for their generosity.

The ideas expressed in this book also poured out of many everyday experiences that wreaked not a few moments of cognitive dissonance in me since the early 1990s, in hindsight a most intellectually stultifying period. Accompanying those moments were occasional conversations and debates that are not easy to have even now, owing to widespread anti-communism. The early 2000s, in contrast, became a nursery for my inchoate, then largely inimical thoughts on socialist states. The fertile turnabout was mainly traceable to the good fortune of becoming part of the now long-defunct New York Editorial Group of *Capitalism Nature Socialism*, in the delightful and enlightening company of the above-thanked Maarten DeKadt, George Martin and Jose Tapia, alongside Paul Bartlett, Paul Cooney, Eliza Darling, Peter Freund, Joel Kovel, Patty Lee Parmalee, Costas Panayotakis and Eddie Yuen. It was during that time that my ideas on state socialism started to turn.

Very gradually those ideas shifted in an unexpected direction. It is a direction that is not necessarily aligned with the thoughts of most, if not all, of those who were part of that New York Editorial Group, and certainly not with mine as they were then. But I still owe all those comrades a profuse thanks for providing the spark, which, under institutional pressures and redirections due to my varied interests, was temporarily extinguished but reignited about a decade later, when I could return to studying the histories of socialism. Eventually, what I managed to publish on the subject intertwined with my background in physical geography and research on soils. The work now before you is to some degree a scion of that interscientific graft, nourished by those who have stimulated me in one way or another, including the Jivi and Baré comrades mentioned in the Preface, to persist in reconsidering and rethinking socialist states in what I hope will be deemed a constructive direction.

1
Introduction

The unpunished disruption of the biosphere by savage and murderous forays on the land and in the air continues. One cannot say too much about the extent to which all these machines that spew fumes spread carnage. Those who have the technological means to find the culprits have no interest in doing so, and those who have an interest in doing so lack the technological means. They have only their intuition and their innermost conviction. We are not against progress, but we do not want progress that is anarchic and criminally neglects the rights of others. We therefore wish to affirm that the battle against the encroachment of the desert is a battle to establish a balance between man, nature, and society. As such it is a political battle above all, and not an act of fate. (Sankara 2007 [1986], 258)

When, in 1983, a popular insurrection reinstated Marxist Pan-Africanist Thomas Isidore Noël Sankara as head of state of Burkina Faso (then called Upper Volta), environmental protection was among the first items in the new revolutionary government's agenda. This complemented commitments to gender equality, public health, literacy and national self-reliance. Those policies were explicitly intertwined in the quest to reach socialism. In the span of four years, until Sankara was assassinated in 1987, mass mobilisations enabled the development of planned wood cutting and livestock movement as well as reforestation to contain desertification (Biney 2018). Since then, with the French neocolonial yoke restored, forests have been diminished from 68,470 to 52,902 km² (as of 2016), a contraction of roughly 23 per cent (World Bank 2021a).

As Burkinabé revolutionaries knew, ecological sustainability is a political struggle and socialism its linchpin. Inheriting and elaborating

1

on a century of socialist experiences and centuries of decolonisation struggles, they had embarked on a promising adventure. They had it right. There is little prospect for an ecologically sustainable society without socialism, without decolonisation. But there was more to this linkage. The Burkinabé revolutionaries showed how socialism is imbued with ecological thinking and how constructive state institutions can be to achieving social equality and environmental sustainability. State socialism, as an intermediate phase, can have and has had ecologically and socially beneficial effects. This can be claimed without downplaying state-socialist problems and horrors. The issue is recognising what has gone well and facing up to what has gone wrong, with an aim to inform current and future efforts, strategies and struggles for an environmentally sustainable, egalitarian, classless, state-free society.

The insights and actions of the Burkinabé revolutionaries could not be more relevant today. A profit-crazed, commerce-glorifying world is the daily stale and toxic bread for most of us. Propelled by the insatiable thirst for profits, capitalist communities are structurally incapable of leaving ecosystems in healthy states. A salient illustration is the settler colonial liberal democracy called Brazil, from where the Bolsonaro regime has intensified encroachments and attacks on Amazonian Indigenous peoples. This is consistent with a long history of attempts to annihilate non-capitalist communities, whose ability to thrive without any need of capital is as intolerable to capitalists as any form of socialism. Socialism, especially in its state-powered variant, may have an uneasy relationship with non-capitalist systems, but with most non-capitalist and anti-capitalist bulwarks gone, there is hardly any restraining the intrinsic rapaciousness of capitalism.

The resulting present is an ever-intensifying concentration and centralisation into fewer hands of the wealth produced by almost all societies worldwide, a systemic tendency of capitalism that Karl Marx underlined long ago (1992 [1867], 776). And there are nauseating repercussions to this. Capitalism is what produces more than a billion people going hungry or malnourished or vulnerable to famines in an age of abundant food production as never witnessed in human history. It is what keeps one in three persons from having access to safe potable water (WHO 2019). It is what gives preferential treatment to housing speculators over those needing shelter. It is the development of profitable and ever deadlier wars, with ever more devastating weaponry, the diffusion of mass imprisonment, mass displacement, mass migration, mass death and vast riches and political privileges for a small fraction of humanity.

The correlate to these preventable, if not politically willed social disasters is the continuing destructive impact on the rest of nature. The capitalist present is a hundredfold speed-up of species extinctions, unparalleled in the history of earth (Ceballos et al. 2017), and an average 68 per cent fall in the populations of mammals, fish, birds, reptiles and amphibians since 1970 (WWF 2020); that is, since many fetters on capitalist activities have been loosened (also known as neoliberalism). The present is over 40 per cent of insect species threatened with extinction (Sánchez-Bayo and Wyckhuys 2019). It is an increasingly more contaminated present that keeps adding to the decades of already mounting quantities of discarded plastics, persistent toxic substances (PCBs, lead, etc.), oil spills, radioactive waste and much poisonous else. The present is a relentless growth in greenhouse gas output into the atmosphere and more frequent extreme weather events (Herring et al. 2020). The present is melting glaciers, rising sea levels, and drowning coastlines and islands. The present is 10 per cent of humanity linked to about 50 per cent of all human-produced CO_2 emissions, a figure roughly mirrored in half of humanity correlating with about 10 per cent of the emitted CO_2. Since 1990, when almost all socialist states were undone, yearly emissions have expanded by 60 per cent, over a third of it related to the lifestyle of the wealthiest 5 per cent (Gore 2020). The present is the gory glory of capital, the free market unleashed, the pinnacle of the magnification of the liberal freedom to loot and kill. That is what really triumphed in 1989. That is what cannot be fathomed by the astonishingly still-existing old believers in capitalism and its hyper-armed and world-policing political correlate, liberal democracy.

Even major capitalist institutions like the World Bank and International Monetary Fund (IMF) acknowledge environmental destruction as a major problem. More importantly, ever more people worldwide understand that the problem is social, not merely technical, and certainly not removed from daily life. Notions like Pachamama, environmental justice, Sumak Kawsay and strike for climate are salient examples of renewed, revived or, in the case of ecologically indebted societies (economically wealthy countries), rediscovered linkages that sometimes cross the large regional wealth gaps dividing the world. These movements recognise and press against the structures of power underlying environmental destruction. In the mainstream, a few in the biophysical and technical sciences appreciate the enormous gaps in political power and the specifically capitalist causes of global environmental destruction (Ceddia 2020; Wiedeman et al. 2020). Regrettably, within those sciences capitalism-friendly prescriptions or technical formulations of regulatory frameworks abound, even when repeatedly phrased in ominous terms (Ripple et al. 2020). The problem is simplistically laid out as one of global population and economic growth. Rarely do technical experts venture into social causes and relations of domination, much less promote an alternative politics, and even less reveal their own political proclivities. When scientists like Wiedeman et al. (2020) feel free enough to speak their mind on capitalism (even naming it) – in a journal like *Nature*, no less – and even point to Marx and socialist alternatives, something major must be afoot. Perhaps in the biophysical or technical sciences people are starting to sense just what a warped understanding of the world most of us have been fed. But even as they seem so close to questioning the foundations of their own society, our brave scientists seem unaware of the huge wealth of historical experiences and resulting ideas, and novel forms of organising, for a society free of domination. Their recommendations spell this all out:

> What is needed are convincing and viable solutions at the systems level that can be followed. We call for the scientific community across all disciplines to identify and support solutions with multi-disciplinary research, for the public to engage in broad discussions about solutions and for policy makers to implement and enable solutions in policy processes. (Wiedeman et al. 2020, 7)

Aside from their ingenuousness on actually existing political processes, where, to name a few problems, Indigenous environmentalists are assassinated and the 'public' is usually met with teargas

4

or worse whenever attempting to be seriously engaged with such issues, these scientists must not know about the existence of Cuba or the notion of the control over the means of production and the not so peaceful struggle that it implies. Nonetheless, under current conditions it would be shocking if these brave scientists are not shunned or accused of 'communism'.

The present is also dotted with stubborn, decentralised, small-area projects worth following, supporting and helping to interlink with each other. There are plenty of them and they point to workable alternatives. Renewable energy and forest conservation increasingly undergird the satisfaction of daily needs in the autonomous Tzotzil Maya egalitarianism-oriented caracoles (municipalities) coordinated through the Zapatista Army of National Liberation (Ejército Zapatista de Liberación Nacional, EZLN) in Chiapas (Mexico). In the centre of the US empire, the Menominee community, in their ongoing struggle for survival as a people, are renowned for their sustainable forestry techniques (Davis 2000). There are many instances of shared urban living arrangements, including squats, that are ecologically less impacting in cities like Barcelona, Melbourne and Rome (Cattaneo and Engel-Di Mauro 2015; Nelson 2018) or in the middle of Virginia in the US (Kinkade and the Twin Oaks Community 2011). Crucially, many Indigenous peoples are on the front lines of defending water, forests, soils and much else worldwide.

Yet these weavings of alternative institutions cannot mobilise people and resources to the level states can, so the myriad anti-capitalist egalitarian projects and efforts, and the movements behind them, have not made the sort of impact needed to diminish substantively the destructive environmental impacts of capitalist systems. Nor should they be expected to carry out such a feat, given the relations of power at play globally. Hundreds of earth defenders, organising resistance against profit-hungry resource looters, are murdered each year (Global Witness 2020). It should not be surprising that solutions cannot be found at the systemic level until systems are radically transformed (i.e. dismantled through replacement), but approaches based on largely uncoordinated small initiatives eschewing state institutions on principle have not proven effective at such transformation. Often, especially where most people are forced into wage dependence, alternative projects cannot offer the means to live well and sometimes just to survive.

There is here an issue of scale. Compared to the relentless and global destructive impacts of capitalist relations, what has so far been achieved – with countless praiseworthy efforts – is manifestly insufficient. This is of course easy to utter, and, to pre-empt any mis-understanding, pointing out the existence of a massive problem is not meant here as any criticism of past and existing alternative projects. The point here is that such projects, crucial as they are, must reach the capacity of overcoming and replacing prevailing institutions worldwide if the aim is to turn things around towards ecologically sustainable social equality. This is another way of saying that a com-mensurate counter-power is needed, implying forms of centralisation (with mutually beneficial divisions of tasks) just to enable coordi-nated global action. It seems that usually alternatives are instead on the defensive, beleaguered and, when getting together, end up repro-ducing pre-existing schisms, if not fragmenting even more. Planetary disasters, like persisting and widening social inequalities, imperialist wars, ozone layer disruption, global warming and ocean acidification and pollution, reflect a net political defeat, at least so far. This should be acknowledged and confronted as much as alternative projects and practices should be highlighted, commended and materially supported. But the task here is not to develop or recommend any guidelines or develop potentially replicable tactics and organising techniques based on successes in bettering the conditions for people and other forms of life. Many have already done so and continue to do so through multiple paths and diverse experiences, as in the examples just mentioned (for more, see Sen 2018). This book is more about recuperating, informing, clarifying and posing questions by drawing from the constructive outcomes of state socialism.

REASONS TO REJECT CAPITALISM AND ITS LIBERAL DEMOCRATIC VARIANT

Capitalism's pairing with indirect or parliamentary democracy is of recent vintage and confined to only a few countries. Capitalism is a set of social relations based on endless, racialised and gendered capital accumulation, private property, national states and liberal to authoritarian ideologies (for instance, 'free-market democracy', nationalism, 'populism', fascism, colonialism and supremacism). Most capitalist societies exist largely as suppliers of commodities for liberal consumption mainly in liberal democratic regimes. They are typically ruled by one or another form of dictatorship, including one-party and monarchical states. For most of its history, capitalism involved strenuous efforts by the ruling classes to deny any form of political participation by majorities. The US, in this regard, is among the stealthiest examples. Were it not for hundreds of years of often violently repressed pressures from below, including from socialist movements, parliamentary democracies with universal suffrage would probably not exist today (Miliband 1994, 24–6).

When parliamentarism and capitalism combine as liberal democracy the situation is not necessarily improved for those living under such regimes, and gets worse for peoples elsewhere by way of imperialism and colonialism. Gerald Horne (1986), among others, have shown how liberalism from its inception is imbued with racism and colonial logic as part of its very class-differentiating foundations (cf. Du Bois 1945). Slavers and 'planters' like Calhoun or Washington cannot be easily dismissed as outside the liberal framework, as if they did not share the same philosophical bases as later figures like J.S. Mill and others of like mind, including those within liberal institutions that hatched fascism and Nazism. After all, Von Mises, Hayek and Croce were all supportive of fascism, at least as a temporary measure to quash worker militancy and save capitalist regimes from themselves. Without liberal democracy and its support for fascists and Nazis as part of anti-communism, fascist and Nazi regimes could not have existed. Liberals of all stripes – conservative, labourite, left liberal, radical, republican, etc. – have much to answer for (Losurdo 2005; Rockhill 2017).

Laconically put, liberal democracy is the art of creating, dumping on and sacrificing others (not just people) to establish or reproduce privileges for a propertied few. From the start, the matter has been sorted out by dividing societies into races. This is why environmental racism is the norm, not the exception, in liberal democracies specifically and capitalist countries generally, whether at the scale of a city or the entire planet. Discussions on and appeals to democracy should always be qualified by appealing to readers' intellectual refinement in distinguishing talk of democracy from support for genocide, slavery and colonialism or from apologetics for, to cite only a couple of examples, the US Confederacy or fascism and Nazism, since those mass-murdering dictatorships were creatures of liberal democratic systems, even elected to parliament (Heller 2011; Perrault 1998).

During the decades when the state-socialist camp was a serious contender to liberal democracies and their allied authoritarian regimes, state-socialist systems were holding in check what is now becoming a much clearer unbridled destruction of working-class life prospects and ransacking of the environment worldwide. It must also be appreciated that the historical achievements of communist, socialist and anarchist movements have been consistently deformed or smashed by liberal democratic forces through constant military pressures and other repressive means (Blackburn 1991, 236; Democratic Socialist Party 1999, 63–4; O'Connor 1998). This is not an excuse for leftist anti-capitalist violence and authoritarianism, but the oppressive turns and, barring anarchism, the statism developing in such movements cannot be understood as if isolated from the societies and international context out of which those movements sprung.

ENVIRONMENTALIST REASONS TO REVISIT STATE SOCIALISM

And yet there is a carefully cultivated and diffused misrepresentation of 'communism' and 'socialism' as metonyms for all sorts of horrors, including the worst environmental carnage the world has ever seen. The European east, in this, has taken on the role of representing the entirety of all things socialist or communist everywhere and for all time. This role is quickly being overshadowed by that of the People's Republic of China (PRC), but the propagandistic effect is virtually the same. As Tickle and Welsh have pointed out:

In the West, eastern Europe [after 1989] quickly became known as a region suffering some of the worst environmental degradation imaginable as a consequence of the excesses of communism.

8

Images from polluted black-spots cast a graphic picture of environmental and human suffering, with little to counter the impression that this was typical of the entire region. The vast tracts of relatively unpolluted lands remained an invisible backdrop to these striking images. (Welsh and Tickle 1998, 17)

Would it look odd to represent democracy and the free market with images of the thousands of Superfund sites in the US, the Cuyahoga River's spontaneous combustion, the Greater London Smog, the countless gouged and barren landscapes from mining operations and heavy metal pollution, the Bhopal pesticide factory explosion, the pulverisation of entire islands with nuclear warheads or the Fukushima meltdown?

True, there were some disastrous environmental impacts through state socialism, but they were neither pervasive nor intrinsic to that form of socialism. In fact, environmental issues were among the priorities even during the Russian Civil War, which followed shortly after the 1917 revolution. Under Bolshevik rule, ecology as a science thrived and became the most advanced in the scientific world. Environmental conservation and movements were part of state-socialist societies, at times shaping national policies if not putting up successful pressures from below (Gare 2002). In several major ways, over a roughly seventy-year period, there certainly were catastrophes, but the net effects were environmentally constructive (see Chapter 4). The following is a brief overview of accomplishments within state-socialist countries that are practicable examples from which ecosocialist futures can be built.

Many species were saved from the brink of extinction or protected by means of large preserves. Preserves were expanded in number and areal extent over time, with some exceptions, and this enabled the protection of entire ecosystems along with their diverse soils and surface waters. To make this possible, millions of people were mobilised to environmental causes and educated formally and informally on the importance of environmental protection, leaving lasting legacies of environmental literacy and sensibility (Goldman 1972; Ostergren and Shvarts 2000; Roman 2018; Rosset and Benjamin 1994; Weiner 1999). Soil conservation measures were, on the whole, successful, given severe economic constraints and inherited low levels of productivity (Betancourt 2020; Brown and Wolf 1984; Chendev et al. 2015; Golosov and Belyaev 2013; Rosset et al. 2011; Wuepper et al. 2020). Logging was kept to within the limits of forest regeneration, for the most part. There were many instances of major afforestation efforts that reduced soil erosion and protected waterways, as well as many species' habitats (Biró et al. 2013; Brain 2011; Potapov et al. 2015; Rosset et al. 2011; Shixiong et al. 2011; Tucker 2000, 48–50). Environmental monitoring programmes were developed in state-socialist countries that industrialised, and monitoring stations, thanks to industrial-level productivity gains, were distributed as densely as possible (Laity 2008; Permitin and Tikunov 1992; Pryde 1991; Whittle and Santos 2006).

In cities, public transportation was privileged over individual motorised vehicle use, which was always highly attenuated. City planning included provisions and actualisations of ample green spaces. Internal travel and internal population migration were restricted, while housing was centrally planned and largely guaranteed, all of which contributed to reducing urban expansion pressures and keeping mobile air pollution sources in check (Goldman 1972; Josephson et al. 2013, 91–3; Koont 2011, 175–6; Pryde 1972). Per capita resource consumption tended to be low and materials recycling was well developed and, at least until the 1980s, highly encouraged (Birman 1989; Gille 2004; Goldman 1972; Krausmann et al. 2016; Peterson 1993, 130). Air and water pollution from industries, though mainly confined to small areas, became problematic especially where state-socialist systems undertook rapid industrialisation, mainly in the 1970s, but measures were adopted that led to substantial improvements in reducing pollutant output and remediating polluted sites. Measures included switching to natural gas where feasible and retrofitting industrial plants with less polluting equipment. In state-socialist countries inheriting a lack of basic sewerage collection and treatment infrastructure, it was thanks to socialist states' introduction of water purification plants that public hygiene was greatly improved (Dominick 1998; Goldman 1972; Muldavin 2000; Placeres et al. 2011; Pryde 1972). Through international treaties, socialist states also promoted biodiversity protection, air pollution and greenhouse gas emissions reduction, and soil conservation worldwide (Josephson et al. 2013, 196–7; Oldfield 2018; Sokolovsky 2004).

Practices and effects were also highly variable historically and geographically. Internal political struggles and external military and economic assaults made environmentally constructive relations extremely tough to achieve or maintain. At the same time, socialist states inherited situations of much environmental devastation and widespread social deprivation. State-socialist countries were also very diverse. A few were already industrialised, but most were largely agrarian and many of those had long been looted by colonising liberal democratic empires. There were certainly major environmentally destructive aspects to state socialism, mainly through the introduction and expansion of increasingly export-oriented mining, fossil fuel production and use, agrochemicals-based and mechanised farming, as well as manufacturing and processing industries. These, however, are typical negatives encountered in any capitalist country and more so in countries reliant on raw material and/or intermediate manufacturing products like state-socialist countries. Unlike the wealthiest capitalist countries, state-socialist countries' sometimes inadequate capacity for pollution prevention or remediation is traceable to an inability to accumulate or direct enough resources to those ends. This is especially understandable when this is viewed relative to the high risk associated with reducing the level of military defence to parry liberal democracies' belligerence. On the whole, socialist states, when not regressing to capitalist ways, as in the PRC, were able to expand or keep natural preserves, contain if not prevent mass species extinction, keep consumption levels low, produce environmentally more sustainable cities than in the most industrialised capitalist countries, expand environmental monitoring capacity and much else, all under immense pressures from within and without throughout their existence. One should draw from and build on such historical experiences and achievements, rather than jettison state socialism entirely.

LEFTIST REASONS TO RECONSIDER STATE SOCIALISM

To those who dismiss or even find offensive any reference to socialism in the 'centrally planned economies', the issue is an obverse of liberal democrats' self-congratulatory and reactionaries' self-privileging delusions. In the case of socialists who deny the socialism part of socialist states, environmental degradation caused through 'really existing socialism' can be easily passed off to one or another form of capitalism or degenerated system that failed to turn socialist. This is a self-absolving purist manoeuvre. In a purist perspective, the notion of triumphant socialism (or communism) is one of an ultimate state of being, where all is resolved and social and ecological harmony prevails. Anything short of that is simply not socialism (or communism).

Many anarchists, on the other hand, converge, though from very different angles, with pro-capitalists in ascribing all sorts of evils as inevitable consequences of state-based or centralised institutions, usually equated with political power-taking or statism. An anarchist position would require taking individual freedom at least as seriously as collective needs and social equality to get to a truly egalitarian society. This is not an impossible ideal; it is lived experience. Copious examples of egalitarianism do exist among non-capitalist societies and even within capitalist ones, but also within state socialism. A recent example is the ABRA Centro Social y Biblioteca Libertaria opened in 2018 in Havana.[1] Anarchists are part of the same societies where revolutions have occurred or where oppression and environmental destruction exist. Anarchists played important, if not prominent roles in socialist revolutions leading to the unintended formation of socialist states. They have not had the level of responsibility as, say, communist parties, for the undesirable or atrocious state-socialist outcomes of revolutions. On the other hand, it is self-serving to claim to have had major roles in revolutions, as in Russia, China and Cuba, while denying connections to any nasty results (cf. Dirlik 1991, 209). Anarchists, too, must face up to the so far consistent failure of achieving the ends for which they have fought and re-examine strategies, as in fact many anarchists have. One instance is the welcome repudiation of propaganda by the deed through bombs and assassinations. Why should the political strategies of communist parties or socialist states not be held to the same standards of understanding

---

1 See https://centrosocialabra.wordpress.com/, accessed 17 January 2021.

and expectations of historical self-transformation, rather than being summarily dismissed?

What is proposed here is a rebuilding by unlearning and relearning about socialist states, though circumscribed to a study of environmental impacts. Just as there are many kinds of capitalism, there have always been diverse socialist currents, among which are Marxism and anarchism. To reject any socialist variant because it is not in accord with one's notions of socialism is to pretend that everyone claiming to be socialist must follow the same line of thinking and action as oneself. This is impossible to achieve even if one tries, and there are anyway fundamentals that are shared and that motivate all socialists, who, in the broadest sense, include anarchists and communists. Stalinists are as Marxist as the early twentieth-century social democrats and the latter-day Trotskyists. The Socialist Revolutionary Fanny Kaplan was as socialist as the Bolshevik Lenin when she tried to assassinate him. Likewise, the anarchists engaged in targeted killings (propaganda by the deed) are just as anarchist as those repudiating such a strategy. Those in the Confederación Nacional del Trabajo who entered the Republican government in Spain during the civil war were just as anarchist as those who rejected that entry in government. Social systems contain the germ of their own supersession, but the seed will not sprout without the warmth and nourishment of self-critical hindsight. It is thus that the horrors and errors of the past are the springboard for better futures.

So, the work here presented is a process of learning from the strides, horrors and mistakes of one general current of socialism, state socialism, from a kaleidoscopic and fractious socialist past and bits of the present, too, rather than consign to a ditch what does not suit one's principles and ideas of socialism, or anarchism or communism. Another reason for this kind of learning may be worth pondering. In a political context where any kind of egalitarian-minded anti-capitalism is immediately equated with historical dustbins or reduced to capitalism-friendly politics (liberalism), it should be a priority to recover or reclaim the results of millions of people's historical struggles for social equality and that, for a time, effectively contained capitalists' power, and curtailed, for decades, their earth-devouring expansionistic tendencies.

There is more to this. The wager here is that there are redeeming qualities to taking the reins of existing politically centralising institutions, if limited to making the coordination of struggles more effective. Socialist states could serve this purpose, or at least it should not be presumed that they are necessarily destructive. There are multiple ways to counter capitalist institutions and to help build power to develop the foundations for an environmentally sustainable egalitarian state-free, classless world. The usefulness of centralising processes depends on what sort of institutions are inherited, the extent to which they can be altered to facilitate egalitarian objectives and prevailing conditions – that is, the results of social struggles. This statement doubtless will turn off not a few leftists immediately, who may justifiably view as an oxymoron this nod to centralism while upholding socialism. This is understandable, but such a stance also involves finding ways to confront aspects of the state that do deliver on people's material well-being and that in some measure legitimise the state in the eyes of majorities. Building parallel institutions that can take over constructive state functions (say, healthcare, environmental monitoring programmes, public education, scientific labs) is important, but this does not necessarily mean that one must take apart or discard everything that has been built. For one thing, securing resources for such parallel institutions is, under current conditions, asking for an immediate clash with the state. This has been historically a losing battle in many ways, and self-defence often entails militarised reorganisation and a quick end to social equality or to equating means and ends. Killing, even in self-defence, becomes part of the regrettable side of a revolutionary process when confronting social forces intent on stamping out any alternative. The current struggle in northern Syria, the Rojava Revolution, is among the many examples.

Still, there are other barriers, as already intimated. It is not just by violent repression that states gain legitimacy. It is people who form the institutions that comprise a state, including millions of people whose lives are economically tied to state institutions (I am one of them). Perhaps too obvious to underline, the state comes out of and is always part of a society, never external to it, so to be against the state is also to struggle with, if not being against a large part of society. Some

14

of this kind of appreciation as opposed to support for the state or for state-like structures is re-entering the leftist arena in writings seeking to demonstrate compatibilities between struggles from within and from outside of state or centralising decision-making institutions (Burbach and Piñeiro 2007; Ciccariello-Maher 2014; Gray 2018). A discussion on political strategies could be more effective when framed in terms of finding complementarities and mutually beneficial coordination. At crucial turns of events, it is helpful to have friends even within oppressive institutions. Besides, relations of domination are more diffuse than the bounds of states. Capitalists, regardless, will ultimately consider all leftists as enemies and try to squash one or the other anti-capitalist leftist variant, depending on which kind of anti-capitalist action is regarded as most menacing for the system as a whole. These thoughts are a subtext of the present work, a not so subtle call to reconsider the socialist state as a viable, maybe even necessary, complement to bottom-up egalitarian action, but these thoughts will not be developed further, save as an open question.

Motivating this book are prevalent misunderstandings and disinformation campaigns about state socialism and socialism generally, the movements that gave rise to socialist states, as well as their environmental records. Given the sustained social and environmental disaster that is capitalism, especially its liberal democratic variant, one should find recurring tropes about socialism (and even more so communism) as being bad for the environment, or of Chernobyl as a metonym for socialism in any form, to be an impediment to the struggle for environmental sustainability. Across political proclivities, many have deemed anything related to 'real existing socialism' peremptorily awful and have celebrated its disappearance. The purpose of this volume is to show that the socialist state has been prematurely discredited.

This may seem outrageous or a futile exercise. After all, 'real existing socialism' was terrible on all accounts. It is a fact, so one is told – something not to bother reckoning or dusting off. But there are facts one is also not told and that can show ways forward from what has been cast aside. A recent study revealed that the only country where people's well-being has been improved in an environmentally sustainable way is in Cuba, arguably the only remaining socialist state (Moran et al. 2008).

Something similarly far from terrible about state socialism was demonstrated in the 1980s. Healthcare professionals have long understood the importance of economic wealth redistribution to people's well-being, such as infant mortality rates, life expectancy and number of people per doctor, hospital and healthcare practitioner, among other such indicators. A study from the 1980s by US-based researchers, using mostly World Bank data, found that state-socialist countries largely surpassed capitalist countries at similar levels of economic development (Cereseto and Waitzkin 1986; Navarro 1992). In Cuba there have been similar public health achievements, bettering even the US when it comes to child mortality rates, in spite of 60 years of sanctions and economic blockade by the US (Fitz 2020). In other words, data gathered through capitalist countries' governments and international capitalist institutions showed how much better off the physical quality of life was or is under socialist states. These are not the delusions of some frothing unreconstructed 'Stalinists' at the fringes of liberal democratic society. They are the conclusions of consummately conventional researchers who take findings seriously.

DEFINING STATE SOCIALISM

Crucial to debates over the environmental impacts of socialist states is defining what is meant by socialist or communist society. Without any clarity on this, one can just plaster environmental ills on whatever social system one dislikes, just to suit one's politics. More importantly, definitions of socialism and communism are 'needed if we are to be clear and persuasive about the kind of future society to which we orient our current activities' (Resnick and Wolff 2002, 75–6). One can start by avoiding literal interpretations and derogatory characterisations. Defining socialism and communism may be highly contested, but there are empirical bounds set through historical experiences and developments that help move this task away from a self-serving game of arbitrary ascriptions (see Chapter 2).

Looking through historical details easily brings to light the uselessness of terms like 'totalitarianism' or 'communist states' to describe countries like Yugoslavia or the USSR. Even conventional scholars had arrived at such conclusion by the 1970s at least (e.g. Jancar 1987, 5–9). Terms like totalitarian, monolithic and the like fail to account for any observable changes within a society and then cannot explain the fall of such systems, especially when the dissolution is instituted by a supposedly totalitarian regime. On the other hand, using official titles to distinguish countries makes for some interesting conclusions, such that the United Kingdom and Saudi Arabia are monarchies and Laos and China are democratic or people's republics. Identifying states or countries or governments as socialist, beyond slurs and labels, is nevertheless fraught with much political contention among more serious activists and scholars, and rightfully so. There has never been a single way of defining socialism. The range has been from technocratic to statist to workerist to decentralist, and so on. It seems, at least, that most on the left would agree that communism means a state-free, egalitarian, classless society, either one worth fighting for or worth drawing from or preserving, if existing state-free and classless societies are taken as akin to 'communist' or 'anarchist' (see, e.g., Baer 2018, 169–71).

Some find it distasteful even to intimate that a state could be socialist because the denotation they use is of a political system defined by bottom-up, democratic decision-making processes, including the workplace (e.g. Chomsky and Pollin 2020, 59–60). This is such a narrow perspective on socialism as to make it impossible to recognise as socialist any government or state ever in existence so far, except maybe for fleeting insurrectionary events like the 1871 Paris Commune or Soviets in 1917 Russia. When doing little more than criticise from a historical or geographical distance one never has to get one's hands dirty, nor assume any responsibilities for anything gone terribly wrong. Furthermore, this attitude also prevents a recognition of positive impacts.

As many of the revolutionary protagonists themselves understood, socialist states were or are in a transitory situation between capitalism and full socialism (D'Mello 2009; Frank 1977; Szalai 2005; Szymanski 1979). It is as structurally a contradictory position as any, but more often a lethal one for many. The processes involved to institute a socialist system can be upended to re-establish capitalism in one or another of its political variants, depending on the outcomes of internal and global dynamics. In the main, a change to a neoliberal variant of capitalism is what has happened. More positively, or as a future prospect, state socialism is a situation that can lead to a fully developed socialist society, with, among other processes, the socialisation of social reproduction and of the means of production, including resource access and production output, systematic redistribution of wealth to cover everyone's daily needs, and with workplace democracy prevailing, all leading to establishing communism, that is, state-free and egalitarian communities. These are not necessary stages, but feasible or desirable ends, dependent on prevailing conditions. The trajectory of a socialist state hinges on the outcomes of both internal relations of power (class struggles, always involving multiple forms of oppression) and external pressures. The latter imply the processes and results of class struggles within countries whose capitalist and/or state institutions bring pressures to bear on socialist states. The reason for framing the matter as involving capitalist and/or state institutions is to keep in mind potential antagonisms among socialist states, independently of capitalist systems, as occurred, for example, between the USSR and PRC and between the PRC and Vietnam. Statecraft is another insidious enemy of socialism, as anarchist communists, autonomists, revolutionary syndicalists and dissident Bolsheviks have long understood.

DEVELOPING CRITERIA

Considering state socialism as transitory or intermediate from capitalism to socialism still does not resolve the problem of identifying criteria to distinguish state socialism from capitalism. Often, socialist states are judged according to who holds political power or how property is handled. Persisting social inequalities in state-socialist societies are explained away through power grabs or usurping control over the means of production. This is behind notions that environmental destruction was due to dictatorship and/or lack of private property (for the pro-capitalists) or workers' control of the economy (for the pro-socialists). But takes like these elide the specific ways in which class formation and conflict occurred in state socialism. It is not possible to explain how state-socialist systems changed without such social dynamism, including how it was possible to have any environmentalist movements emerge, as they did.

Blaikie (1985, 34), in trying to build a comparative framework to explain differences in socially caused soil erosion rates in different countries, drew attention to the importance of considering the relations of production affecting land use. He surmised that relations of production in 'centrally planned economies' have specific characteristics, but unfortunately did not elaborate much further on this point. Resnick and Wolff (2002), paying attention to the concept of mode of production (how production is organised in a society), provide a possible answer. They distinguish between property relations (ownership, control of the means of production), political power and surplus labour processes (production, appropriation and distribution of surplus; production relations). The latter (what happens in the workplace and how what is produced is distributed in society) characterises class structure. On this basis, they define communism as a form of government (a state) addressing 'class structure' (workplace dynamics, vying for classlessness and the disposal of the surplus produced). Communism is when producers of surplus are the same as the people disposing of the surplus; hence, workers' cooperatives could be an illustration of this. Socialism overlaps with communism (though it is unclear how) but does not tackle 'class structure'; that is, socialism is about 'state management, regulation, and intervention in the economy (perhaps including nationalization of productive property, planning, and so on) to secure greater equality of incomes, a broad social welfare minimum, mass democratic political participation, and so on' (Resnick and Wolff 2002, 77).

But relations in the workplace and how surplus is taken and distributed are just as consequential to class differentiation as who owns what and who has what sort of political power. This is because the relative control over the means of production is part of what constitutes one's class position. Among the reasons why a manager can tell me what to do at work (production relations) is, among other reasons, because that manager has been granted control over an office or a factory floor by a boss or owner. The reason that surplus can be distributed in ways contrary to most workers' benefit has a lot to do with different forms of political power (not just in the state), and the extent to which workers can exert such power through formal or informal social institutions. Class structure is defined so narrowly by Resnick and Wolff as to downplay or deny the effects on class differentiation of who owns what and who has what sort of political influence. The processes might be separable analytically, but not in lived experience, and focusing exclusively on the employer–employee relationship obscures substantive differences between capitalist and socialist states. Socialist states, among other things, were systems where necessary and surplus labour were decided politically through centralised state organs, not through highly variable market pricing conditions dictated by capitalists (Cockshott and Cottrell 1993). Wealth redistribution through variously planned investment allocations was hardly a way for capitalists (or bureaucrats or party members) to get rich. If they were just another form of capitalism, one would need to explain the need to dismantle such systems to privatise an already available source of capital accumulation, the insistence on Marxism-Leninism, the virtually guaranteed employment, the free social services and the maintenance of low wealth gaps.

A promising clarification on the dynamics of a mode of production turns out to be vitiated by a reductionistic take on class structure and a confused if not tautological terminology. In Resnick and Wolf's understanding, capitalist class structure is defined by capitalist forms of surplus appropriation. The class structure is the employer–employee relationship, which defines the ways surplus is made, taken and dished out, which creates the capitalist class structure. What is more, contrary to Marx's view, Resnick and Wolf reckon that 'communism' has a class structure and even a state and government policy to boot, including the possibility of autocracy. That can be one way of understanding communism, but it is not the meaning adopted in this volume. Their definition just adds to the reigning confusion on communism. Startlingly, Resnick and Wolff omit state-free societies that are or have been classless. There are even examples under their nose, in the US (the settler colonial system where those authors live) in the everyday practices and social institutions of traditionalists in 'Reservations', for example. The departure from not just Marx, but communists and even anarchists, about the meaning of communism and socialism is no mere semantic quibble. Confining class relations to matters of surplus is what allows Resnick and Wolff to claim socialist states as state capitalist (whether socialism or communism has ever existed anywhere, the authors never tell us). Essentially, countries like the USSR were never socialist because they retained the employer–employee relationship in the workplace. This takes us back to the purism that exonerates socialism from any environmental destruction and at the same time disallows acknowledgement of positive results.

Resnick and Wolff's analytical differentiation of political power, property relations and surplus labour and management are still useful, but need further specification. Coming to the rescue are James O'Connor (1998, 258–62) and David Lane (2014, 7), who provide handy, concise descriptions for state socialism that do not efface substantive differences.[2] Lane takes socialist movements' histories in much greater consideration by defining state socialism as:

> a society distinguished by a state-owned, more or less centrally administered economy controlled by a dominant communist party which seeks, on the basis of Marxism-Leninist ideology and through the agency of the state, to mobilize the population to make a classless society [i.e. communism].

This definition includes a system-specific set of property relations and political power structure, including a superstructure component (e.g. ideology), whose importance should never be underestimated. O'Connor emphasises many of the features encapsulated in Lane's pithier description, but importantly adds social policy aspects of commitment to socio-economic equality, full employment, job security guarantees and collective consumption (such as public transit, common eateries, cooperative farms, etc.), as well as commonalities in the social conditions and economic status of the places where state socialism emerged.

---

2   See also Baer (2018, 46-50) and similar, though carelessly worded, criteria in Jancar (1987, 13–15).

Lane's claim that 'State socialist societies were insulated from the world capitalist system' (2014, 8) is best set aside. Viewed biophysically, the assertion of state-socialist insulation from the rest of the world implies the ability of existing on another planet. Three examples should put the matter to rest. One is long-range air pollution bringing contaminants across different regions. Another example is stratospheric ozone layer-thinning chemicals emitted mainly from the largest capitalist economies but affecting many other parts of the world, especially at higher latitudes. Finally, and by now more obviously, greenhouse gas emissions, also overwhelmingly emitted from the largest capitalist economies, have led to radiative forcing in the atmosphere (global warming, climate change), affecting differentially the entire planet. There is no insulation possible from such environmental impacts.

The insulationist view also implies negligible to no trade, scientific exchanges and much else, a most mistaken assumption contravened by a cursory look at the evidence (O'Connor 1998, 258–9). For example, the USSR and their allied states were already in a dependent relationship with western European and North American countries by the mid-1950s, trading mainly raw materials for key manufactures like machinery and other commodities tied to technological innovations. It was an increasingly disadvantageous relationship, where North American and western European powers could restrict the kinds of technologies exported to state-socialist countries to maintain economic and military superiority. As economies reliant on raw materials exports, state-socialist countries were also vulnerable to rapid downturns due to poor terms of trade. For instance, when oil and sugar prices took a dive in the 1980s, the USSR and Cuban economies suffered major revenue shortfalls. The economic dependency on liberal democracies, as the world's largest consumer markets and centres of highest capital accumulation and technological innovation, only worsened over time through high-interest loans, joint companies and direct investments, among other forms of capitalist encroachment (Berend 1996; Böröcz 1992; Sanchez-Sibony 2014).

What is missing in Lane's definition is the issue of class structure, which existed and exists in state socialism, and that Resnick and Wolff have the great merit of raising as a main parameter. However, class structure in state-socialist countries resulted mainly from overt political struggles, rather than concealment through market exchange. The state was the main locus of decision-making processes over the degree of surplus production and allocation. That was predicated on differential control over the means of production, a degree of control that varied over time as a result of class struggles, expressed within the ruling party as well. The centralised appropriation and redistribution of surplus, largely run by older men often from a predominant nationality or cultural group, occurred in uneven gendered (largely patriarchal) and ethnically based if not racist ways, as the Roma and some Siberian peoples experienced.

Still more refined criteria are needed for state socialism than Lane develops. The above description implies that currently China, Cuba, Laos and Vietnam are all state-socialist. This is true insofar as those states retain institutional structures characterised by, for example, the channelling of political involvement through a single party, social assistance programmes and reliance on 'Marxist-Leninist' terminology for political legitimation. If material conditions and historical developments are taken as a primary way of defining state socialism, those socialist states are no longer socialist because much of the economy is in the hands of private businesses and ideological justifications are now permeated by neoliberal terminology, among other reasons. This is one among many issues to be clarified, with attention to historical conjunctures.

To Lane's description several other attributes can be added for greater clarity and for more encompassing societal comparisons. One essential attribute of a socialist state is its formation out of self-defined communist movements ousting a capitalism-oriented regime, whether a semi-feudal monarchy, a settler colonial or colonial dictatorship, a neocolonial 'democracy', etc. Historically there has not yet been any successful socialist revolution toppling core capitalist regimes, but this does not mean it is an impossibility. The corollary of taking down a capitalist regime is at least a formal intent, including once in power, of fostering, primarily through state institutions, the development of the conditions for a communist society. Whether steps were or are actually taken to achieve such a goal is something that should be explained rather than expected or dismissed as impossible. What counts as a step towards communism also depends on the official and/or assumed meanings of communism taken within the countries affected. This is one reason, for instance, to categorise as state-socialist the present political systems in place in Cuba, Laos, the PRC and Vietnam, even as the Cuban case departs substantively from the others and is, arguably, the closest to the socialist path. The idea of the PRC or Vietnam as state-socialist will likely be rejected by many for differing reasons, but the implications of ascribing the capitalism label to such countries are actually counterproductive to socialist (and especially ecosocialist) causes (see the second section in Chapter 4, 'The PRC: Fulcrum of World Ecosocialist Struggles').

Another essential attribute is relative to social chronology. State-socialist systems have been established where states already existed. This qualification is important, given the existence of other kinds of civilisations featuring similarities to socialism or communism but that never developed a state as a main way of structuring society (or that historically decentralised and became state-free, but later came under the yoke of a state). Just as important, this sequential specificity to the establishment of state socialism brings into relief major influential factors that are typically ignored, which are the persistent structural inequalities engendered through inherited state institutions, among other relations of domination. These alone take multiple generations to overcome and, arguably, cannot even be neutralised when surrounded by belligerent national states intent on the destruction of any form of socialism (understood as anti-capitalist in orientation). Anarchists in the Russian and Spanish civil wars, to allude to some salient examples, learned this the hard way, so the challenge is not confined to socialist states (Graham 2002). It was and will be a challenge to face for as long as capitalist societies exist.

These criteria (Table 1.1) serve as the main way of categorising with greater precision the state-based social systems described as socialist by their governments or ruling parties. A list of state-socialist countries is provided in Table 1.2 with their respective periods of existence. According to the above-discussed criteria, 26 countries were state-socialist and only one, Cuba, remains today. Notably, relative to environmental impacts, 13 never industrialised under socialist states and most of them had short time spans. This is based on

Table 1.1 Main criteria to identify countries as state-socialist

Historical and institutional processes	Principal characteristics
Wider societal context of revolution	Capitalist or capitalism-oriented system
Power-taking revolutionary formation	Self-defined communist party or movement
Originally stated political objectives	Building socialism to create conditions for communism
Formal political system	Dominance of or exclusionary rule by a communist party
Class structure	Based on politically centralised surplus production, appropriation and distribution
Official ideology	'Marxism-Leninism'; state-centred communist party leadership
Organisation of main economic activities	Centrally administered or coordinated
Property relations in key economic sectors	Predominance of state ownership
Social policies	Socio-economic equality; full employment and job security; collective resource consumption

a study by Bentzen et al. (2013), except that allegedly deindustrialised countries are still counted here as industrialised because they retain an industrial infrastructure for most economic activities, including consumption. Accordingly, twelve had a pre-existing industrial base, though for most of them it was under state socialism that industrialisation really expanded to national levels. Of these, the central European countries, North Korea and the western parts of Russia inherited major environmental problems by the time they turned state-socialist.

Three countries, while retaining nominally communist party rule, 'Marxism-Leninism' and control over some key economic sectors, have embarked on an overt capitalist path, and mainly industrialised after that major shift. Not a few self-described socialist systems with one-party rule are therefore not regarded as socialist states, such as Laos, the PRC and Vietnam. This may raise eyebrows. However,

Table 1.2   State-socialist countries

Country	Period	Years	Substantive industrialisation relative to state socialism	
			Prior	During
Afghanistan, Democratic Republic	1978–92	12	No	No
Albania, People's Socialist Republic	1946–92	46	No	Yes
Angola, People's Republic	1975–92	17	No	No
Benin, People's Republic	1975–90	15	No	No
Bulgaria, People's Republic	1946–90	44	No	Yes
Burkina Faso	1984–7	4	No	No
China, People's Republic	1949–78	29	No	Yes
Congo, People's Republic	1969–92	23	No	No
Cuba, Republic	1960–	60	No	Yes
Czechoslovak Socialist Republic	1948–90	42	Yes	Yes
Ethiopia, People's Democratic Republic	1974–91	17	No	No
Germany, Democratic Republic	1949–89	40	Yes	Yes
Grenada, People's Revolutionary Government	1979–83	4	No	No
Hungary, People's Republic	1949–89	40	Yes	Yes
Kampuchea, People's Republic	1975–91	16	No	No
Korea, Democratic People's Republic	1948–1974	26	Yes	Yes
Lao People's Democratic Republic	1975–86	11	No	No
Madagascar, Democratic Republic	1975–92	17	No	No
Mongolia, People's Republic	1924–92	68	No	Yes
Mozambique, People's Republic	1975–90	15	No	No
Polish People's Republic	1945–89	44	Yes	Yes
Romania, Socialist Republic	1947–89	42	No	Yes
Union of Soviet Socialist Republics	1922–91	69	Yes	Yes
Vietnam, Socialist Republic	1945–86	41	No	No
Yemen, People's Democratic Republic	1969–90	23	No	No
Yugoslavia, Socialist Federal Republic	1945–91	46	No	Yes

Sources: Modified from Bentzen et al. (2013) and https://en.wikipedia.org/wiki/List_of_socialist_states (accessed 14 January 2021).

the reforms of 1978 in the PRC and the similar policy shifts in Laos and Vietnam in 1986 spelled a decisive turn towards capitalism in those countries. State ownership lost predominance in most sectors and state firms can fail under capitalist competitive pressures. Mass unemployment problems ensued and have been absorbed mostly through a constant expansion in output by other firms or in other sectors, including through foreign direct investment. Key sectors like manufacturing and farming are under private control and much economic activity is coordinated through private domestic and foreign enterprises. Class structure is mainly based on relations of surplus production and appropriation dominated by private firms with surplus distribution mostly achieved through capitalist labour market pricing mechanisms. In other words, in those countries there is clearly a capitalist predominance in the relations of production and in the management of surplus labour, but there remain key socialist aspects. An example is a dominant coordinating role of the state and an official Marxist-Leninist ideology. Extensive public services are provided, and socialist forms of property are not only legally recognised, but in part still favoured through government policies (on China, see Long et al. 2018).

The above shifts towards a capitalist road resemble the development of other formal instances of 'socialism' in other countries, where a one-party socialist government presides over a capitalist economy. Such governments (twelve in total) demarcated or demarcate themselves emphatically away from the sort of systems espoused in the USSR or Maoist China. There is no communist objective in such governments and Marxism-Leninism was mostly or entirely rejected. State involvement in economic activities have varied historically by degrees, as in liberal democracies. A list of these countries is given in Table 1.3.

Self-titled socialist administrations have ranged from one-party governments, such as the Baathists in Iraq and Syria, to multi-party governments under predominant socialist party influence, as in Nicaragua with the Sandinistas. Other examples are Egypt under the Arab Socialist Union and Tanzania under the Tanganyika African National Union. The Sandinista government in Nicaragua explicitly followed principles of political pluralism and a mixed economy.

Table 1.3   Socialist governments with capitalist economies

Country	Period	Years	Substantive industrialisation relative to one-party socialist government	
			Prior	During
Bolivia	2005–19; 2020–	14	No	No
Cape Verde	1975–92	17	No	No
China, People's Republic	1978–	42	Yes	Yes
Eritrea	1991–	29	No	No
Iraq	1958–2003	45	No	Yes
Lao People's Democratic Republic	1986–	34	No	No
Libya	1969–2011	42	No	Yes
Myanmar	1962–88	26	No	No
Nicaragua	1979–90	11	No	No
Somalia	1969–91	22	No	No
Sudan	1969–85	16	No	No
Syria	1963–2011	48	No	No
Tanzania	1961–85	24	No	No
Venezuela	1998–	22	Yes	Yes
Vietnam, Socialist Republic	1986–	34	No	Yes

Sources: Modified from Bentzen et al. (2013) and https://en.wikipedia.org/wiki/List_of_socialist_states (accessed 14 January 2021).

Many of these alternative socialist governments were corporatist: more focused on national sovereignty and building national identities. This was often as a way of ensuring the sheer existence of new states carved out of former colonies, within neocolonial frameworks, as in the case of post-1968 Iraq in relation to the US (the US government helped Saddam Hussein rise to power and sustained his regime militarily through the 1980s). The substance of their economies and internal and foreign relations were highly contingent on political developments within liberal democracies like the US, France, Italy and the UK.

The kinds of environmental policies (if they had the possibility of establishing any) and impacts such one-party socialist governments had were heavily shaped by direct meddling from capitalist countries, mainly by means of neocolonial relations, including proxy and direct wars. This is why it is important to distinguish socialist states, like the USSR or the pre-1986 Lao People's Democratic Republic, from socialist governments with capitalist economies or under neocolonial influence, like the Syrian Arab Republic or Burma under the Burma Socialist Programme Party. It is not just a matter of relative sovereignty, but also of what ultimate and shorter-term political objectives guided the state. This distinction, it turns out, has environmental ramifications, largely on account of the PRC in terms of the severity of environmental impacts since the Dèng reforms.

STATE SOCIALISM AND THE ENVIRONMENT

The following chapters proceed first, in Chapter 2, with a historical and geographical overview of the rise and development of socialist movements and states. This background is important in dispelling pervasive mischaracterisations of socialism and socialist states. Exposing and dissecting such ideas about state socialism is taken up in Chapter 3 by means of empirical comparisons that demonstrate the mixed to positive environmental impacts of state socialism and by refuting the logic of comparisons that fail to consider global inter-linkages. Chapter 4 provides the wider and specific contexts that need to be addressed to explain the particular forms of state-socialist environmental impacts. This is accomplished by discussing the USSR, PRC and Cuba. Chapter 5 is dedicated to explaining the relationship between state socialism and the environment, first by examining and critiquing more popular perspectives and then formulating an alternative historical and dialectical materialist explanation. Chapter 5 is a discussion of and set of questions on the contradictory nature of state socialism relative to building an environmentally sustainable egalitarian classless and state-free society.

Chapter 2 is a highly abridged history and geography of socialisms. Just like other political movements and ideas, socialism has never been in the singular. This has been so from its very beginnings. So, whenever anyone reviles or embraces socialism, the first question one should ask is: which socialism? Throughout I highlight ecological worldviews expressed in socialist currents. This is to dispel prevailing false notions about socialism, including about ecological understandings, and the oxymoron of the idea of communist states, a concept that fundamentally contradicts what all communist movements and thinkers have stood for. The latter part of the chapter is dedicated to describing the recent emergence of ecosocialist movements and thought.

Chapter 3 takes up the popular claim that the environmental records of state-socialist countries are the worst ever or worse than those of capitalist democracies. Such views are predicated on tendentious sensationalised examples and arbitrary comparisons. Criteria or analytical frameworks are typically not spelled out and verdicts are dispensed mainly on the basis of impressions or selective use of evidence, rather than substantiated by systematic inquiry. The frameworks of the comparisons are usually implicit and often based on faulty assumptions. They can be absolutist (the worst), synchronous and diachronous. When comparisons are done thoroughly and a diversity of types of evidence is considered, it is evident that the reality is mostly the reverse of what we are told. Changes over time in the same countries (diachronous analyses) are found to be more appropriate, though insufficient, in catching more comparable trends through similar ecosystems. Overall, capitalist systems, especially liberal democracies, are much worse for the planet, historically and in the present. This is even more so if one includes the internationally profitable business made out of the Chinese economy since the 1980s. Ultimately, comparing countries that are historically interconnected and that have mutually changed each other is a dubious undertaking. On biophysical grounds, comparisons are for the most part problematic and much more research is needed to gain a clearer understanding of how to compare different biophysical contexts and histories. For these reasons, comparisons should at the very least be contextualised in a country's social and environmental history and according to global processes and interlinkages across countries, all of which vary over time.

Going beyond debunking prevailing views, Chapter 4 delves into the general processes shared by state-socialist countries and the various conditions where socialist states arose. The task is to consider the historical, international and biophysical contexts. This includes giving attention to the legacies of capitalist impacts on the environments where socialist states were established and to a fuller record of socialist states beyond the usual fixation on negative impacts. The task is carried out through studies of the USSR, the PRC and Cuba. Each case study features a discussion of what went well as well as badly and why, while bringing in comparisons with similar environmental impacts in liberal democracies, especially the US.

Having covered empirical evidence, varying contexts and major social and environmental changes, the discussion moves to appraising explanations and developing a dialectical and historical materialist framework in Chapter 5, the concluding chapter. On biophysical degradation problems in state-socialist countries there has been no shortage of theories. The general argument is that socialist states are intrinsically destructive biophysically because they are economically inefficient authoritarian systems suppressing the counterbalancing nature of civil society, censoring crucial information, mismanaging resource use, chronically compromised by institutionalised conflicts of interests and lacking private property and market signals. These claims are unflinchingly made in the face of similarly destructive tendencies in liberal democracies and are contradicted by much of the evidence presented. By and large, such explanations amount to little more than celebrations of one or another idealised form of capitalism. Analyses tend to be vitiated by evaluations based on whether or how well socialist states conform to standards allegedly characterising the wealthiest liberal democratic countries. Most explanatory statements therefore end up being a mere reproduction of preconceived notions. Recent, alternative 'revisionist' frameworks, in contrast, consider the wider context and local particularities of socialist states and appreciate the importance of the variety of biophysical processes. These scholars look for deeper processes that thread through state-socialist and capitalist systems alike. Underlining this convergence, they look for causal mechanisms common to both kinds of social systems. However, far from explaining anything, these arguments confuse convergence with sameness and are incapable of sorting out what set socialist states apart as well as identifying and explaining relations of power internationally and subnationally that led to biophysical degradation or improvements.

Explanations from leftist circles find some similarities with the convergence theorists, but also avoid being mired in allegedly greater depth or nuance. These theories are largely inchoate and so are followed by an alternative explanatory take modelled after the author's prior research, building on existing leftist theories and approaches more broadly. This attempt at a dialectical materialist framework is sketched out mainly as an outline, as otherwise a separate book could (perhaps should) be written. Putting these thoughts into print may understandably invite heavy criticism if not rebuke from at least part of the left. Rest assured, for what it is worth, that I do not find socialist states necessarily preferable to liberal democracies. This depends on the trajectory that emerges out of struggles within socialist states relative to the objective of reaching an environmentally sustainable socialism. The main problem, for those like me living in capitalist countries, is the uses of state socialism to scare folks away from socialism or communism in general and to divert attention away from the historical and current horrors of capitalist societies, whether in their liberal democratic or authoritarian garbs. A corollary problem is the treatment of state socialism as irredeemable, as if entirely devoid of positive potentials through struggles from within.

Chapter 5's conclusion is an opening to the implications of the findings for ecosocialist strategy. A main argument is that rejecting state socialism is premature or not the most constructive route, considering how state-socialist systems helped reduce or mitigate the destructive tendencies of capitalism and given the deterioration of social and environmental conditions worldwide, especially since the 1990s. What should be done instead is to revisit and learn from socialist states so as to build on their strengths and overcome or pre-empt their awful aspects. One way to begin doing so is to discern the causal factors common to socialist states that led them to some environmentally destructive episodes. Several broad interrelated causal factors are traceable to the wider context of a capitalist world economy; the interconnections between socialist states and capitalist powers, if not former colonial powers; and pre-existing and newly formed internal social conflicts, including class struggles. These three processes combine to create overarching contradictions that to a large extent will have to be faced by any current socialist formation: building the foundations for a future classless and state-free society and the defences to fend off capitalist powers and survive within a capitalist world economy, while bringing sufficient material well-being to all and less (not more) environmental harm. Addressing these contradictions involves two forms of interrelated but different struggles: a social one and a biophysical one. Socialist states, for all their flaws, exemplified this combined struggle. They still offer much not only in terms of signposts about what to prevent, but also of potentials to overcome capitalist relations in environmentally sustainable ways.

This work departs in several ways from the copious writings on environmental degradation and socialist states. First, it is a comparative approach to state-socialist systems' environmental impacts at different scales of analysis. This allows for greater attentiveness to environmental processes, which are not bound to political boundaries. Second, state-socialist systems and biophysical processes are studied as dynamic and changing over time and as mutually constitutive. The terms environmental and biophysical are used interchangeably. Biophysical as understood here has two subsets: ecological (relations among organisms and between organisms and physical environments) and physical (solar radiation, wind, wave action, etc.). Finally, there is emphasis on a relational, multiple-scale analysis of environmental impacts within state-socialist countries. Relational here means examining world-scale linkages and mutual influences between state-socialist and capitalist countries as well as considering global biophysical changes to help explain biophysical effects in state-socialist countries. This comparative, relational and multiple-scale framework that includes the study of biophysical processes is a longer-winded way of saying historical and dialectical materialism, an often mischaracterised and shunned way of understanding and explaining. More importantly, adopting this framework is a response to the usual ways scholars and pundits treat socialist states and environments. Socialist states are often depicted as monolithic, unchanging and stagnant, and environments as passive substrate or background. Overall this work addresses state socialism in relation to capitalism on several fronts, political and scientific.

2

A Brief History of Socialist States and Ecosocialism

Traceable in places to social struggles, philosophies and beliefs in the middle of the 1600s, socialism has always been in the plural, a polyvocal response and set of alternatives to the social and environmental degradation and havoc wreaked by capitalism. Its multifariousness is rooted in diverse and historically shifting peasant, artisanal and proletarian social conditions and value systems (Arrighi et al. 1989). Even if largely emerging out of capitalist European societies, socialism was inflected from the start by the influences of communalistic and egalitarian paragons elsewhere, especially from Indigenous peoples' lifeways in the Americas and revolutionary self-determination movements, as through the 1791–1804 Haitian Revolution (Anderson 2010; Birchall 1997; Krader 1974, 5–6; Meisenhelder 1995; Merchant 1980; Weaver 2014, 276).

Practices, strategies and ideals already varied greatly by the time socialism emerged as a concept in the 1820s. Some currents, very few at first, were inclusive and sensitive to multiple forms of oppression, and some were masculinist, chauvinistic and even justified colonialism and racism. There were those, such as Cabet, Fourier and Owen, who imagined and at times set up egalitarian communities or workers' cooperatives that, more often than not, fell apart within a few years. Some of these communities were religious in character, basing their socialist quest on dissenting interpretations of holy scriptures. There were those advocating for technocratic centralist rule by producers, including industrialists, whose foremost exponent was Saint Simon. His followers pressed the state for the establishment of a technocratic order based on the large-scale scientific management of the economy.

Syndicalists focused on worker power as the way to reach socialism and formed unions to that effect. The visionary Inka-French feminist

Flora Tristán introduced the concept and practices of working-class unity in the struggle to improve workers' conditions at both work and home (Beik and Beik 1993). Most syndicalists, however, largely accepted the legitimacy of the existing social order, at least in economically prosperous years. Finally, Jacobins, a prominent grouping during the French revolution, strove to turn state institutions into organs for the gradual extension of political rights to all. Shared by all the major socialist movements was the belief that everyone involved in production could eventually cooperate to improve society for mutual benefit, which assumed a harmonious relationship between, for instance, peasants, industrialists and workers (Cole 1953).

The delineation of the now familiar labels of socialist, communist and anarchist took roughly a century. Anarchism and communism would start assuming clearly separable politics in the 1860s, largely splitting not too long thereafter. A demarcation between socialism and communism would not appear until the 1920s splits within socialist and social democratic parties (Braunthal 1967a; Graham 2005; Guérin 1970).

In the 1830s, with the spread of workers' mutual-assistance associations, Pierre Joseph Proudhon developed the concept of anarchism as an alternative kind of social order based on mutualism and federated communities. These ideas would be further elaborated on and diffused by Mihail Bakunin, Élisée Reclus, Pëtr Kropotkin and the Owenite William Benbouw, among others, who stressed the inseparability of state and capitalism and therefore the necessity to overcome both simultaneously. Imbued with these ideas, revolutionary syndicalist and anarcho-communist movements organised revolts in places like Ukraine, Mexico and Spain in the 1900s (Eckhardt 2016; Graham 2005; Guérin 1970). At the same time, Kropotkin and especially the communard Reclus, whose work is an early form of bioregionalism, started to call attention to the importance of the environment in a non-deterministic fashion (Reclus 2013).

Karl Marx's and Friedrich Engels' analyses of capitalism as a system of dispossession and exploitation and their development of dialectical historical materialism started maturing in the 1840s. The *Communist Manifesto* (1948), written for the Communist League (1847–52) and revived in the 1870s, was an early distillation of this framework. It was part of a new approach eventually called Scientific Socialism, grounded on studying social relations, both past and present, so as to arrive at political strategies attuned to existing possibilities. Such an overarching worldview offered an alternative to a tendency in prior and contemporary perspectives, deemed Utopian, to formulate pre-packaged blueprints devoid of analyses of social context and hence unable to find effective applicability, attested by negligible political effects. Noteworthy are Engels' early studies involving attention to environmental degradation and Marx's scattered, laconic, but clear analyses of the environmentally destructive propensities of capitalism (Dickens 1992; Foster 2000).

Neither Scientific Socialism nor historical and dialectical materialism were ever unified perspectives under any one theoretical or party line. Much effort would be exerted from the very beginning, especially by Friedrich Engels, to counter various non-dialectical, economically deterministic (vulgar materialist) and idealist perspectives claiming scientific standing. The struggle for scientific rigour and dialectical materialist open-endedness was ultimately unsuccessful, as rigid, closed and sometimes doctrinaire worldviews have often prevailed within Marxism, institutional and otherwise.

REVOLUTIONS AND STATE POWER

In much of Europe and, partly, in European settler colonies, the ideas of Marx and Engels came to prominence especially through the International Workingmen's Association (the First International, 1864–76), where they played key roles, alongside collectivist anarchists like Bakunin. The First International was, among other things, a focal point of decisive political bifurcation among socialist forces. Engels, Marx and their allies (including a few communards) tried to centralise decision-making processes under their direction and favour the formation of parties to be coordinated by means of the International. Anarchists and some syndicalists balked at any state-oriented route, while Bakuninists formed shadow and conspiratorial groups. The schism that resulted from the increasing divide, exacerbated by state repression, fractured the First International irreparably, but its demise was overshadowed by the rise of social democratic parties (Braunthal 1967a, 175; Cole 1961; Eckhardt 2016).

As a result of recurring brutal repression, preceding and during the First International's existence, some socialists favoured forming secret societies to achieve their aims, by force if necessary. This route became particularly easier to justify after the 1871 massacre and persecution of thousands of communards by the Prussian and French governments to squash the Paris Commune. Other tendencies saw social change unfolding through tensions within society itself and opted for gradualist approaches, working within existing institutions. Some socialist formations developed both kinds of strategies, or shifted from one to the other. This diversity of strategies has run across all socialist currents at one point or another.

From the late 1800s, socialism was progressively represented by parliamentary or statist approaches. The founding of the Second International in 1889 was dominated by political parties, especially the German Social Democratic Party. In 1893, anarchist formations and trade unions were made officially ineligible for membership, leading to a permanent severance of anarchist organisations from what became mainstream socialist institutions. Out of these changes also came greater affinity between socialist and labour parties and statism, and then reformism. Anarchists and syndicalists would thenceforth form their own independent international organisations and rarely interact with other socialist groups (Braunthal 1967b).

Communism was integral to Marxism early on but also to some forms of anarchism, where the term was inflected as anarchist communism. The differences from Marxism lay mainly in seeing the state as a fundamental obstacle to be overcome immediately and in refusing any intermediate stage to achieving communism. Marx and Engels, in contrast, upheld strategies of movement centralisation, predicated on striving for working-class gains, and the conquest and use of the state so as to dissolve it. They defined the process as proceeding in two steps: first establishing worker majority rule ('dictatorship of the proletariat'), and then, through that transitional stage, the eventual achievement of communism. State socialism, arguably, conforms to this notion of a transitory phase.

Marx and Engels would later see a potential for skipping directly to communism in non-capitalist societies, but this would be contingent on successful revolutions in the most powerful capitalist countries (Rodney 2011). Those writings would not be widely known until the 1960s. The means to achieve communism – a state-free, classless society – would nevertheless be highly contested within Marxism, ranging from conforming to and/or using liberal democracy to a violent overthrow of any form of capitalist regime and its replacement with a socialist version (Cole 1960; Graham 2005). Be that as it may, for both Marx and Engels, as for Reclus, in different ways, comprehending biophysical processes (a 'mastery of nature') was crucial to developing a communist society. They warned against what would now be called environmentally unsustainable practices. These proclivities persisted within Marxist thought (e.g. Rosa Luxemburg, Vladimir Ilyich Ulyanov 'Lenin'). They furnished precedents and continuities that would form socialist states' environmentalist undercurrents, nourished by histories of struggles against colonial depredations, the material effects of industrialisation and recurring and, mainly by the 1980s, rediscoveries of insights from Marx and Engels (Dickens 1992; Gare 1993; Löwy 2017).

In contrast, some socialist, social democratic and labour parties adopted narrow and gradualist objectives and, in some respects, even liberal principles, confining political work mainly to parliamentarism. In the early 1900s, anti-Marxist and Marxist revisionist currents arose that converged over a disavowal of class struggle. Communism remained the province of still-influential segments of social democratic parties. In the case of the Russian Social Democratic and Labour Party (founded in 1898), such a segment was embodied by what came to be called the Bolshevik wing, who split from the Mensheviks by 1912 and founded their own party.

The Bolsheviks were ahead of their time. The contradictions within social democratic parties became a centrifugal force that, in the wake of World War I, tore them apart on questions of nationalism and support for war. Most social democratic parties (including labour and socialist parties) opted on the side of their respective national governments in a crude display of chauvinism, flatly contradicting their 1907 anti-militarism resolution. A minority within the

Second International, convened at the Bolshevik Lenin's insistence, met in 1915 to organise opposition to the war, but it was ultimately in vain. Nationalist binges in mass slaughter prevailed upon the Second International, which was officially defunct by 1916.

Excepting Russia, revolutionary attempts like the 1919 German Spartacist revolt and Hungarian Soviet Republic were crushed militarily within months. Newly formed or reconstituted authoritarian regimes subsequently incarcerated, executed or exiled thousands of socialists of all stripes. Similar convulsions in other parts of the world preceded or roughly coincided with the 1917 Russian Revolution, even if not as influential at the time. In Mexico, the Mexican Revolution of 1910–20 included anarchism-leaning Magonismo and the pro-peasant agrarian socialist Zapatismo. The institutions that came about would later enable Cuban revolutionaries to set up their main base in Mexico in the 1950s. In China, the republic founded with the overthrow of the monarchy (the 1911 Xinhai Revolution) encouraged the spread of socialist ideas. These infused the 1919 May Fourth Movement, which also had anarchist currents and included activists, including Máo Zédōng, who founded the Communist Party of China in 1921 and became their leader in 1943. Otherwise, out of the mayhem of World War I and schisms within most social democratic and socialist parties, leftist currents often broke away to form communist parties, either as the war was ending or shortly thereafter.

In Russia, a militarily weakened Czarist dictatorship gave way, in October 1917, to the rising tides of workers' councils (soviets) occupying factories and mines, peasant communes and land appropriations, national liberation movements (especially in Siberia and Central Asia) and a variety of socialist formations, mainly the Bolsheviks, Mensheviks, Social Revolutionaries and, to some degree, anarcho-communists (e.g. Makhno-led revolutionaries in Ukraine). The Bolsheviks succeeded not only in gaining popularity in the industrialised, urban and especially western parts of Russia, but also in centralising decision-making processes and consolidating their own political power. This they achieved by providing workable alternatives enabling many to survive the wars, repressing other socialist formations and crushing revolts, and eviscerating soviets of decision-making authority.[1] Developing a new army (the Red Army), the Bolsheviks, with the aid of Left Socialist Revolutionaries and Anarchists (Makhnovists), eventually, by 1922, defeated multiple, well-armed military formations (the White Guard) allied to the Czarists. Notably, the feat included the repulsion of many invading foreign troops supporting the White Guard, including thousands sent by the US Wilson government (the American Expeditionary Force, Siberia). These liberal democratic and subsequent Nazi-fascist invasions in World War II would permanently scar and define the politics of the new socialist state, the USSR. A major outcome was the privileging of military investment to ensure self-defence, the rapid industrialisation that would provide such military capacity and the contradictory environmental practices and ramifications that would follow.

THE RUSSIAN REVOLUTION, THE SOCIALIST STATE AND THE ENVIRONMENT

Communist fortunes changed radically with the Russian Revolution, for the better for some and for the worst for others. One-party Bolshevik rule was resisted or denounced by dissident communists and anarchist communists from the very beginning. There were nevertheless major social gains and a flowering of egalitarian culture. It was during the 1920s that ecologists, naturalists and conservationists exerted increasing influence and major strides were made in environmental protection (Gare 1993; Weiner 1999). These promising developments were curtailed by the late 1930s, under Stalin's leadership, as the aftermath of internecine struggle among Bolshevik factions over economic policies, political strategies and the level and kind of centralisation. Political purges occurred multiple times and thousands of people were disappeared or interned. Several million perished or suffered greatly, while millions more had their lives improved and a small percentage accrued prestige and power.

---

1   For this reason I invite the reader to rethink the acronym USSR as the Union of Socialist States centred about Russia. This is because soviets were rapidly suppressed after the revolution, Bolshevik rule emanated overwhelmingly from Moscow and yet the stated objectives and policies remained socialist throughout.

The repression created even more critics from within as well as without. Prominent among them was Trotsky, founder of the Red Army. Organising actions outside the party to counter the rising influence of Stalin's faction, he was expelled from the party in 1927. Living in exile, he founded the Fourth International in 1938 as response to the Third or Communist International (the Comintern, 1919–43). He was assassinated in Mexico by Stalin's agents in 1940. Critiques of the USSR under Stalin would vary with time and according to political persuasion, but largely centred on the lack of substantive worker control over the economy and the suppression of legitimate dissent. It is from these perspectives that the first critical studies of socialist states emerged. Developing and applying Marxist theories, Trotsky and his allies concluded that the USSR and similar societies had become degenerated workers' states, with later Trotsky- ists going further in regarding such social systems as bureaucratically deformed, ruled by a new and increasingly entrenched bureaucratic class, or, in other renditions, as state capitalist (Chase-Dunn 1982). Criticism was not solely about politics, though. It included environ- mental issues. But such criticism was largely indirect and confined to scientific communities and naturalist organisations (Weiner 1999).

Within two decades Russia was being transformed forcefully, fever- ishly and violently into an industrialised society, formally organised into federated socialist republics under the rule of what came to be called the Communist Party of the Soviet Union (CPSU). Decision-making was centralised in a restricted number of people, often the CPSU's Politburo, sweeping aside the early Bolshevik internal policy of 'free- dom of discussion, unity of action' (democratic centralism). Forced requisitions and subordination to farm managers (euphemistically called collectivisation) meant that the peasant majority effectively lost control over much of the land appropriated. Mixtures of economic coercion and enticement induced ever greater employment in facto- ries and cooperative farms. While it became an increasingly repressive country for many Bolsheviks as well, the USSR quickly began to rival economically and militarily the most advanced industrialised capital- ist countries. This proved crucial to gaining popular consent, to curry leftists' support abroad, as well as to rout the Nazi and imperial Japa- nese invaders from the west and east during World War II.

Many communists and socialists drew great inspiration from the Russian Revolution and the Bolsheviks in particular. They began forming communist parties and/or aligning themselves with the Bolsheviks through the Moscow-centred Comintern. There, communist and allied organisations would meet periodically and coordinate activities, at least in the first years, to promote global proletarian revolution. Decisions made through the Comintern came to have far-reaching consequences, some of them entirely unintended.

For example, strategies for the colonies and relative to oppressed peoples were to involve the cultivation of solidarity among workers across different communities and alliances with liberal elements in national liberation movements. And the various affected parties dutifully aligned their activities accordingly in their respective countries. This is what led to the early collaboration between the Communist Party of China (CPC) and the Guómíndǎng (the Nationalist Party) until, after the death of the party founder, Sun Yat-sen (Sun Zhongshan), the Guómíndǎng purged and slaughtered communists in 1927. The reverberations were, among other places, felt by the communist anti-colonial Thanh Nien movement, which folded as a result of losing their primary base of operations in China and their main leader Nguyen Ai Quoc (Ho Chí Minh), who escaped to the USSR.

From 1928 to 1935 the Comintern called on communist parties to undermine moderate leftist parties. Thereafter, with the rise of Nazism, the policy shifted to a popular front strategy, characterised by forming broad alliances to fight fascism. Such abrupt shifts created confusion, demoralised or marginalised activists, and created conditions conducive to conformism and personal vendettas. Worker-led revolutions, as in Spain (1936), were sacrificed in the name of protecting the USSR, or socialism in one country, in a sense making a universalising socialist virtue out of the particularity of Bolshevik self-preservation necessity.

The repercussions of Bolshevism deeply affected many leftist movements in other parts of the world. Those allied to or taking inspiration from the Bolsheviks turned ever more towards the statist and vanguardist approaches, forming the ideological backbone of state-socialist systems. The objectives and techniques, to simplify, amounted to founding and expanding a rigidly disciplined, militarised party structure, often in clandestine form (mimicking experiences in Czarist Russia, where such strategies were essential for sheer survival). The next steps entailed building and severing alliances as necessary to conquer state institutions by violent means, involving infiltration into military institutions or at a minimum the formation of a well-trained military wing. Conquest of state apparatuses would be followed by the consolidation of power and elimination of opposition groups, including those of the leftist variety. The party, claiming to represent the workers, would then embark upon proletarianisation (including subduing self-subsisting communities) and industrialisation, repressing or smashing any movement getting in the way. This was justified on the basis of creating the conditions for socialism, understood as a preliminary stage towards communism.

As the established regime bureaucratised and militarised (and institutional Marxism-Leninism ossified into a set of dogmas and precepts), forced industrialisation became an attractive way in many countries, especially former colonies, to ensure economic dependence by most people on state organs (the consolidation of power) as well as to raise material well-being (the cultivation of popular consent) in ways that, ironically, converged with aspects of contemporary liberal democratic notions of modernisation. Industrialisation also proved of great value in self-defence against the constant threats of invasion or coups by liberal democracies or other capitalist states. The matter of self-defence may have been useful for propaganda purposes, especially in marshalling nationalist sentiment, but it was also a legitimate preoccupation. It must be recalled, for example, that the US invaded Russia in 1918–20 and China in 1900–1 to try and suffocate revolutionary movements and that such policies as military build-up and alliances like the Warsaw Pact, founded in 1955, were responses to global US military expansionism and to the 1948 formation and subsequent expansion of NATO (Berend 1996).

SOCIALIST STATES IN A CAPITALIST WORLD ECONOMY

What was done in the name of workers and socialism in the USSR was perceived as successful by many leftists, who replicated or attempted to replicate Bolshevik strategies elsewhere. The very existence of the USSR had also greatly modified the world economy and the political contexts of countries where leftist forces were emergent. At the same time, new socialist states established through Bolshevik-modelled and supported revolutions were expected to be subordinated to the CPSU Politburo.

To some extent, such an arrangement was established with communist parties in western Europe, but more in terms of coordination, which did not last much beyond the early 1970s. By and large actual subordination occurred in a few central and eastern European countries. This unfolded through the USSR Red Army's liberation from Nazi and fascist regimes and subsequent military presence, which pre-empted political reversals in favour of the enemy camp and stifled independent socialist state action. Salient examples of the latter are the military suppression of the mainly communist revolts in Hungary in 1956 and in Czechoslovakia in 1968 (Berend 1996).

The degree of the USSR's grip over 'satellite' states was always tenuous, though. The Red Army withdrew in 1958 from Romania, where the government pursued an independent path, essentially disengaging from the Warsaw Pact by 1963 and becoming among the US government's 'Most Favoured Nation' trading partners in 1975. In 1973, after the capitalism-leaning reforms of 1968, the Hungarian government acceded to the General Agreement on Tariffs and Trade (the precursor of the World Trade Organisation), and then the IMF in 1983. These illustrate a progressive subordination to the core capitalist bloc during the USSR's existence (Berend 1996).

Nor was there the sort of neocolonial dependence rife among formerly colonised societies. By the late 1970s the USSR was basically subsidising its supposed 'satellites' to the tune of $20 and $80 billion through an unequal exchange of raw materials and energy for manufactures (Turnock 2006, 284). There were also socialist states forged out of autochthonous forces expelling the fascist invaders, like Albania and Yugoslavia. Those countries were always independent of and repeatedly clashed with the USSR. Overall, in environmental matters, as well as domestic social policy and even internal government politics (which could even involve multiple parties, as in the

German Democratic Republic), socialist states were independent of the USSR (Berend 1996).

The 1949 founding of the PRC played a major role in shifting revolutionary leadership away from Europe, as it eventuated into rivalry with the USSR over political hegemony among communist formations worldwide. The schism happened by the late 1950s, leading to an undeclared border war in 1969 and rapprochement with the US government, made obvious with US President Nixon's 1972 visit. There was also much internal strife within the PRC, for reasons not too dissimilar from those in the USSR during the 1930s. A major manifestation of the conflicts was the tumultuous and deadly Chinese Cultural Revolution (1966–76). The result, the ousting of the Máo faction, eventually yielded a reformed CPC under Dèng Xiǎopíng's leadership (Li 2016; Xu 2018).

Almost all ruling socialist formations (socialist states or one-party socialist governments) were formed between the 1960s and 1970s, during or after the USSR–PRC schism and the independent lines taken by Yugoslavia and Albania. Moreover, all such ruling socialist formations were grounded in anti-colonial national liberationist movements and ideologies and deeply etched by long colonial histories of racialised segregation. That there would be conflicts among ruling socialist formations within and between socialist countries should therefore not be surprising. For example, in 1979 Vietnam and the PRC were at war in relation to the Vietnamese government, with USSR support, invading Cambodia to oust the Khmer Rouge, which had historically developed out of the Vietnamese-dominated Communist Party of Indochina, founded in 1930 by Hồ Chí Minh (also known as Nguyễn Sinh Cung).[2]

---

2 The Khmer Rouge had removed the US-backed Lon Nol dictatorship after the US carpet bombed the country (1965–73) and had killed hundreds of thousands of civilians (with the excuse of destroying Việt Nam Cộng-sản military corridors). In power the Khmer Rouge, composed mainly of non-communists, engaged in even greater mass atrocities as they robbed an already starved people of even the basic means of survival, tortured and murdered thousands, and imposed forced labour to accumulate capital for eventual industrialisation (Tyner 2017; Vltchek 2015, 610–11). Following liberation with the crucial assistance of state-socialist Vietnam, dissident Khmer Rouge took over the government and, after more than a decade of one-party rule, eventually became the currently dominant Cambodia People's Party, in the framework of a representative democracy and constitutional monarchy. The ousted Khmer

The situation and trajectories elsewhere could therefore not be more divergent from those in Europe, and the USSR had even less influence in such places. Revolutionary struggles were deeply intertwined with decolonisation and involved alliances with nationalist and liberal democratic elements. Harsh colonial repression induced secretive organising, making Bolshevik strategies more attractive and effective. In colonies and semi-colonies in north Africa and much of Asia, Marxist social democratic parties had already formed since the days of the Second International and then Bolshevik parties by the 1920s. Most of them were repressed or essentially destroyed by governments aligned with core capitalist states.

On one occasion, North Korea's socialist rule degenerated into a dynastic autocracy. In the wake of civilian-decimating and infrastructure-flattening US aggression and following prolonged power struggles within the Workers' Party of Korea, the national-chauvinistic Kim Il Sung Kapsan faction seized power. They steered the country away from socialism by 1972 in favour of an amalgam of voluntarism, nationalism and neo-Confucianism, known as Juche or 'self-reliance' ideology (David-West 2011; Gills 1992; Robinson 2007).

In much of Africa, when independence movements successfully overthrew colonial rule, mainly in the 1960s and 1970s, the timing largely coincided with established socialist state systems in conflict. This affected newly independent states in terms of fleeting linkages and support, as the USSR and PRC supported either right- or left-wing governments depending on geopolitical convenience. This was not a one-way process. Excepting the marginalised Pan-Africanist and Afrocentric Communist movements, who promoted African forms of collectivism, self-described socialist or communist parties were rather fluid politically, aligning themselves with liberal democracies, the USSR or the PRC according to changing circumstances.

To some extent, what transpired internationally largely reflected the societal diversity negated by boundaries and states imposed by colonisers and the widely different histories of African communities.

Rouge regrouped, took refuge in Thailand, and had the institutional support of both the US and Chinese governments until the 1990s. The Vietnamese army would withdraw only by 1989.

Many African peoples had no backgrounds of highly centralised states or sometimes any state formation at all. They had also undergone little to no industrialisation, so that Bolshevism had arguably greater difficulty spreading compared to places in Africa affected by settler colonialism and industrialisation such as South Africa. Pre-existing communalistic institutions in many Asian and African societies could be used towards developing and diffusing socialist ideas or practices, but the capacity for self-subsistence in most peasant, pastoralist and gatherer-hunter communities also often made socialist notions redundant. Hence, a diffuse radicalised social base, supporting otherwise well-organised political groups, was often absent and this also impeded any firm and lasting political footing in most countries. A consequence was a mix of socialist and liberal policies characterising most socialist governments and states. The Somali single-party Barré administration is a paragon of this, switching sides according to which government offered a better military and economic deal.

Furthermore, socialist states or one-party socialist governments in African countries, but also in many Asian countries, could not emulate Bolshevik strategies because they lacked the economic and military means to do so. When gaining state power, communist and socialist parties, much like their counterparts in contemporary socialist states (by the 1970s), were easily undercut through raw material pricing (uneven terms of trade), direct financial pressures (e.g. extortionate interests on loans) and other such tactics from core capitalist countries or former colonial powers, which were often the same.

Otherwise, as in the cases of Angola and Mozambique, the new socialist states were under constant military attack and invasions by the South African racist state, supported by the US and allies, or by proxies of former colonial powers. Other countries suffered a fate akin to that of the above-described, short-lived Burkinabé socialist state (1983–7), which was struck down through a murderous pro-French coup before any lasting positive social and environmental impact could even start to materialise. Such horrific post-liberation conditions made it virtually impossible to carry out any sensible environmental policy or to start reversing the long-term biophysical damage of colonial dictatorship. Excluding the relatively ephemeral

Burkinabé case, the average lifespan of socialist states in Africa was 17.3 years, barely enough time to leave much of a mark. Rather, the environmentally ruinous mass extraction of raw materials to benefit core and semi-peripheral capitalist economies continued virtually unabated through neocolonial relations, facilitated through supranational institutions like the IMF and World Bank.

The Americas and Asia-Pacific, dominated by settler colonists through racist systems, have been rough terrain for working-class unity of action through national liberation and decolonisation. Aside from draconian policies to separate people according to racial categories, the annihilation of many Indigenous and Aboriginal peoples and the appropriation of their lands necessitated an entirely different approach to socialism, one that never really came about as a result of racism or Eurocentrism among settler colonial socialists (Bedford and Irving-Stephens 2000). The various early Utopian socialist communities formed in the US by the likes of Cabet and Owen in the 1800s attest to a widespread indifference among socialists of the day to the racism, genocide and settler colonial land theft that enabled the establishment of such Utopian socialist communities. The revolution in Haiti (1791–1804) – a yoke-shattering, world-transforming freedom struggle – was and still is little appreciated in socialist movements (Blackburn 2006). Environmental impacts in the Americas are thus overwhelmingly tied to shifts in various types of capitalist regimes.

Yet hundreds of years of Indigenous and Afro-descendent peoples' liberation struggles have nurtured movements that have led to some socialist successes in taking over the reins of government. The pluri-national state of Bolivia and the Bolivarian state of Venezuela are the main illustrations of socialist rule over a still capitalist economy. Mixed-heritage peasant movements have also played a major role, as in the establishment of the Sandinista government in Nicaragua. But these are short-lived, limited or very recent turns of events. A veritable contrast is Cuba. Since sloughing off US colonial dominance in 1959, Cuba exemplifies continuity despite huge odds. There has been repression, not too unlike in other state-socialist contexts, but never to mass-murdering proportions. Arguably state socialism became a default pathway out of neocolonial stagnation. More is said about Cuba in Chapter 4, but the environmental repercussions have eventually been the making of the most environmentally sustainable country on the planet.

Aside from various forms of often repressive if not mass-murdering statism and misguided vanguardism, there were also traces of Saint-Simonianism in the self-congratulatory importance given to parties as omniscient guides or incarnations of the proletariat. This streak would explode into an increasingly dominant technocratic wing in socialist states as they came to be increasingly integrated into the economic orbits of core capitalist countries, mainly those of western Europe and North America. The intensifying economic dependence of socialist states on capital from liberal democracies translated politically into the reform phase of the 1960s and 1970s (Frank 1989). During this period, party affiliation, especially in central and east European state-socialist countries, started to become less economically important (though still politically prestigious) than technocratic status. A dual system came into temporary existence characterised by military and political dependence on the USSR and economic dependence on the most powerful capitalist states (Böröcz 1992).

The PRC would also partially succumb to such internal party friction, but in different ways and through deadlier struggles (i.e. the Cultural Revolution, 1966–76) eventuating in the Dèng reforms of 1978 and the more or less fully fledged capitalism in the present under CPC direction, with key economic sectors under direct state control. By the 1980s, other socialist states would follow suit in South East Asia (e.g. Vietnam's 1986 Đổi Mới reforms), as they would in the USSR and through much of the state-socialist world. Huberman and Sweezy (1968, 117–20) summed up well some of the major contradictory characteristics of the USSR and with it arguably most other state-socialist countries when they deemed the system highly stratified, 'effectively depoliticized' and capable, by means of private rather than socially oriented incentives, of rivalling the likes of Japan in coaxing ever more productivity from workers. Still, the histories and development of socialist states and their environmental impacts are incomprehensible without factoring in internal struggles, capitalist legacies, pressures from belligerent capitalist states and changing social and environmental conditions (see also Chase-Dunn 1982).

ECOSOCIALISM

Socialist states may be the centre of attention in this work, but, as briefly discussed, socialism has always included movements resisting or refusing centralisation or state power. I return to these only to highlight those that have influenced the development of ecosocialist ideas and organising. This is because such socialist movements give primary status to environmental concerns.

Ecosocialism, briefly put, is a movement, perspective and by now even an institutional politics that gathers socialist and environmentalist principles and objectives together. It is socialist in the sense of identifying capitalist relations as the ultimate and systemic cause of structural inequalities and environmental destruction. Politically, this means struggling for social equality by establishing the social control of the means to life. This includes decolonisation and cross-generational justice as well as overcoming patriarchal relations and developing respect for differing knowledge systems, striving to combine them to the benefit of all. Ecosocialism is environmentalist in calling attention to the biophysically destructive character of currently conventional ways of living and in premising the understanding of biophysical processes on diverse forms of systematic knowledge and inquiry, institutional and otherwise. Ecosocialism, in other words, stands for the development of biophysically sustainable egalitarian communities worldwide (Kovel 2014; Löwy 2011; Turner and Brownhill 2006).

Like socialism has been historically, ecosocialism is just as divided regarding political strategies, ways of organising and other such matters. Hans Baer (2018, 136–53) provides a useful overview of the different currents. A salient difference is among those who see state power or centralisation as essential, those who find such strategy anathema and those who strive to find complementarity between grassroots and state institutions.

To some extent, as remarked above, environmental concern was already expressed in the writings of Marx, Engels, Reclus, Kropotkin, Luxemburg and Lenin, among others, but it was not really until the 1960s that socialist movements returned to and elaborated on the germinal ideas within the socialisms of the 1800s and early 1900s

(Foster 2000; Gare 1993). The global 1968 revolts were a particularly important juncture in this recovery and elaboration process.

Anarchist-communist, feminist, Maoist and Trotskyist organisations were influential in those revolts, an outpouring of liberation movements long excluded or marginalised within most socialist or state-socialist institutions. In many countries, including in liberal democracies like Mexico, such movements suffered from police brutality, assassinations and outright massacres during protests. In the US, one may recall the Black Panther Party, among other national liberation movements like the Young Lords and the American Indian Movement, who were methodically and ruthlessly squashed through liberal democratic state repression (Shawki 2006).

Such organisations, if not the ideas they promoted, have been important in the formation and development of existing anti-systemic movements, through the World Social Forum and other international groupings like Via Campesina. The latter was constituted largely by smallholding farmers and has aided the development of low-input farming in Cuba. There are also revolutionary movements and communities drawing inspiration directly from communist and socialist histories and ideas, but on the basis of their complementarities with locally specific egalitarian communalistic traditions, political and philosophical thought and current social conditions. A few salient examples of such anti-authoritarian forces at the time of writing include the Brazilian Movimento dos Trabalhadores Rurais Sem Terra (Landless Workers Movement, formed in 1984), the South African Landless People's Movement (founded in 2001), the largely Tzotzil Mayan EZLN (founded in 1983) and the militarily besieged and mostly Kurdish Democratic Confederalism of Rojava (Northern Syria, established in 2011). In core capitalist countries there are or have been similar movements like Occupy, the Anarchist Black Cross and Co-operative Jackson in North America and various communities in Italy struggling to re-establish the commons (e.g. NoTav in the Piedmont region), as well as long-standing squatters movements in many metropolitan areas (Akuno and Nangwaya 2017; Cattaneo and Engel-Di Mauro 2015). This is besides the continuity of relatively small political formations and periodically erupting popular demands for more state-socialist kinds of provisions, like guaranteed employment, workplace rights, unemployment benefits, health care and much else.

There have been socialist ideas infused to some extent in environmentalist movements as well, especially since the late 1980s. Several theorist-activists and dissidents in state-socialist countries began to reassess and, in many cases, prematurely or opportunistically reject Marxist approaches from an environmental standpoint, and in some cases to mine classical works for ideas, until then mostly ignored, on people–environment relations. Arguably, more advanced environmental understanding in socialism or communism emerged in areas of the world where struggles for self-determination or sheer survival (e.g. decolonisation) involve the protection of ecosystems such as forests. Hence, the Brazilian Seringueiros movement led by Chico Mendes (assassinated in 1988) was among the first to combine ecological and socialist approaches in an explicit manner by the early to middle 1980s (Löwy 2011). The relatively late attendance to ecological thought within socialist and communist movements (and later in some parties, too) is only partly explainable by the predominance of socialist state industrialisation prerogatives, in their varied garbs. The preoccupations of the majority of socialists and communists lay firmly within the social. After all, it was the social question that lay at centre stage in the very origins of socialism. Furthermore, socialists and communists were not immune to the society–nature dualism and other ideological constructs typical of capitalist societies.

Historically it has therefore been challenging to overcome such predominant worldviews. The 1990s, however, saw the bridging of what have been often called red and green perspectives, congealed in emerging and spreading red-green movements. In some case, green parties, for example, have incorporated some traditional socialist issues as part of their platforms, as in the UK. Conversely, anarchist-communist, Trotskyist and other communist parties and organisations have increasingly adopted environmental issues as their own, such as in the Fourth International. Efforts have also continued to be made to draw red-green movements closer to the numerous struggles of Indigenous peoples, especially as a result of the latter's worldviews and everyday practices being traditionally more constructive or holistic relative to nature. Such major self-critical renewals are

also behind the development of what have become known as ecosocialist perspectives. Aside from movements taking up these ideas,
including Kovel and Löwy's *Ecosocialist Manifesto* (2001), there have
been several political formations and even state institutions where
ecosocialism is increasingly being incorporated in platforms and
policies, as in the Bolivian and Venezuelan governments (the latter
even having a ministry dedicated to ecosocialism), the Left-Green
Movement (Iceland), the Nordic Green Left Alliance and the Partido
Socialismo e Libertade (Brazil). What is also novel to these developments in socialist and communist movements and institutional
political formations is their attentiveness to and promotion of gender,
anti-racist and decolonial egalitarian outlooks, viewed as crucial to
ecosocialist transformation (Baer 2018).

With this historical overview and multiple-scale relations of power
in mind, Chapter 3 addresses and takes apart prevailing ways in
which state-socialist and capitalist systems are compared relative to
environmental records. First, conventional methodology is applied
to disprove the view, based on that same methodology, that environmental impacts are worse under state socialism. Afterwards, the
conventional methodology is subjected to critique on account of the
explanatory interlinkages and contexts it erases, and a different comparative analysis is developed.

3

The Poverty of Comparisons

To put the point as bluntly as possible, socialist revolution has
proven to be less ecologically harmful than capitalist imperialist
rivalry and counterrevolution.

James O'Connor (1998, 257)

An exceedingly warm day in late May 2020 in northern Siberia is all
it took to liquefy the permafrost land supports of a diesel fuel tank at
a thermal power plant near Norilsk, among the cities most polluted
by heavy metals and sulphur emissions. The structure collapsed due
to its own weight and 6,000 tonnes of diesel seeped into the soil.
Another 15,000 tonnes reached nearby waterways and the turbid,
grey-brown water of the Ambarnaya River turned crimson. It will
take decades for this part of the Arctic to recover ecologically. So far,
in Russia, only the 1994 Komi crude oil spill surpasses the amount
spread by the Norilsk disaster. Yet this major spill pales in comparison to the Exxon Valdez disaster near Tatitlek, Alaska. In March 1989,
with the beginning of the end of European state socialism, an Exxon
tanker smashed into a reef at Prince William Sound, releasing some
37,000 tonnes of crude oil, immediately ravaging surrounding ocean
ecosystems and another 2,100 km of coastline. And yet this is still
a small fraction of the all-time record-setting Deepwater Horizon
explosion at the hands of BP. The oil released and still being released
has polluted about 1,770 km of coastline. The Norilsk disaster is
therefore not the worst by any means. It is, however, the latest in a
long and ever more sordid record of fossil fuel havoc wreaked on
oceans, seas and inland waters and on all beings drawing their suste-
nance directly therefrom. The taiga and broad-leaf coniferous forests
in the Ural region have been contaminated with crude oil since the
USSR was made to vanish (Buzmakov et al. 2019). The Norilsk diesel

spill, along with all the largest, high-frequency and low-magnitude fossil fuel spills, is a crystal clear free-market product, since the raw material is extracted by a private company for profit, to be sold in domestic and world markets.

Not that Norilsk – home of the world's largest smelting complex – had previously been a centre of ecological harmony. Very far from it. Under the USSR, starting in the 1930s with forced labour, Norilsk eventually joined the list of most polluted cities in the world, with its concentration of smelters and mining operations. The Kola Peninsula is peppered with pockets of lasting damage, most intensified since the 1970s, including taiga forests diminished by acid rain downwind of nickel smelters and piles of radioactive waste from nuclear plants (Bruno 2016; Kozlov and Barcan 2000; Revich 1995). Other cities, like Chelyabinsk and Magnitogorsk, bear the environmental scars of similar policies. These cases, often and rightfully decried as monstrosities, were modelled after industrial centres in the US, like Gary, Indiana (Josephson et al. 2013, 84), among the most enduringly polluted places (Dietrich et al. 2019; Hurley 1988). It should be little surprise that shifting to capitalism has not at all improved on prior disasters and certainly not on the USSR's conservation achievements. The diesel spill is a testament to worsening conditions in other aspects of environmental impacts. Yet continuing and in some ways worsening pollution problems are glossed over when the USSR's environmental record is evaluated. And the same logic is extended to socialism generally. Typical arguments begin with some assertion that state socialism was (or is) worse on the environment than those of capitalist 'democratic' societies.

Often the comparison (even when parenthetical) ends there. Socialism is just categorically worse to the environment. Full stop. It is common sense. Supporting evidence is optional. And it is for good reason that such totalising statements are rarely backed by data. Because when comparisons are made at all, the results end up embarrassing purveyors of free-market democracy. Some conditions under state socialism are similar to or even better than under industrialised liberal democracies, like the relatively much lower levels of consumption and waste (either total or per capita; see Birman 1989; Goldman 1972; Krausmann et al. 2016). Other aspects may be worse but in delimited zones, like air quality around Norilsk or sulphur emissions from coal-fired plants and heating systems.

Another instance of using environmental problems to defame socialism is the infamous 'black triangle', where air quality still tends to be poor. It is a roughly Belgium-sized heavily industrialised area located at the meeting points of the Czech Republic, the former German Democratic Republic, and Poland. But this comparison is hardly unequivocal when compared to the frequent deadly smog enveloping the Po Valley in northern Italy or the areas in and around Delhi, in northern India, or the recurring toxic air in the same, now formerly state-socialist 'black triangle'. And Los Angeles had its fair share of smog problems for decades before heavy regulations, not more free-market measures or more 'democracy', were put in place.

It might seem simple enough to parry any such critique with constant reminders of Chernobyl, as if it says everything about what socialism meant for the environment and people. A more critical view of cases like the US's Three Mile Island disaster should temper those kinds of pronouncements, as Marshall Goldman (1988) would have reminded us. One could create an equally false image of the US, Canada and other liberal democracies by showing only the most devastated areas in those countries. Here are a few examples: the destruction of the Florida Everglades, the pollution of the Niger Delta, the Bhopal disaster, the radioactive releases at Hanford in the US, the more than 3,000 highly contaminated Superfund sites in the US and acid deposition in western and central Europe.

It seems that comparisons are more often rhetorical window dressing than attempts to get to grips at all with the causes of environmental destruction. Most of the landscapes in state-socialist countries have not been negatively impacted, if impacted at all, compared to what was done prior to the political change that brought about socialist states. This could and should be said of China now as well. In any case, there is a prevailing omission about similarities. To draw from Philip Pryde: 'history must deem the 1970s and 1980s as decades of net environmental losses. This is equally true in both the United States and the Soviet Union, where striking parallels exist in the context of environmental problems' (Pryde 1991, 291). In other words, the environmental records of industrialised state-socialist and liberal democratic countries are more a case of convergence, not differential ranking, but – it is important to add – for divergent causes created through interlinkages across state-socialist and capitalist countries.

Anti-socialist environmentalism is like the shallow, fig-leaf variety ('econationalism') expressed by many protest groups in the latter-day USSR, where environmental devastation continued or recrudesced almost as soon as local bosses could grab for themselves a piece of the USSR and call it independence. Comparisons are usually not followed through. Perhaps it is because they do not lead unequivocally to capitalist superiority.

Comparative statements are largely deployed as pre-emptive intervention to foreclose debate. And in terms of physical health (Cereseto and Waitzkin 1986; Navarro 1992), capitalist detractors of state socialism would certainly lose the argument. This is among the open secrets that gets very little attention and very rare mention. But on the environment, comparisons can get easily muddled because it is not as straightforward as pitting some measure of economic development against physical health indicators. There are different ecosystems with widely different kinds of species and characteristics, multiple biophysical combinations, long-term and shifting weather patterns, and so much else to consider. The variability is quite the head spinner. What is being compared is a much greater challenge than is usually recognised. But when comparisons of environmental impact are indeed made between capitalism and state socialism, they tend to be self-servingly selective, supporting the pre-packaged conclusion that free-market democracy is best.

MAKING COMPARATIVE FRAMEWORKS EXPLICIT

As already discussed about such selectivity, why are most countries ignored, and why are the criteria for comparison seldom specified? What is supposed to be compared to what and on what basis? In the case of socialist states, it seems more appropriate to compare industrialising countries with similar countries or to compare according to comparable levels of economic status or timing of industrialisation. This would be more sensible than pretending that, say, much

of western Europe, with a couple of hundred years of industrialisation, is comparable to south-eastern Europe (e.g. Albania, Bulgaria,
Romania), with very recent industrialisation packed into a few
decades. After all, many environmentalists pin the blame on industrialisation for our current environmental woes (as if technology, not
politics, were the malaise to be addressed). It also makes little sense
to compare Japan or West Germany with Hungary, Poland or Yugoslavia (instead of, say, Czechoslovakia and the German Democratic
Republic), because manufacturing and mechanisation, among other
such developments, largely occurred from the 1960s onwards in those
state-socialist countries, not, as in Japan, since the late nineteenth
century. South Korea or Taiwan may make more sense to compare to
Hungary and Poland, in this respect. In any case, if one is keen on the
environment, ecosystem comparisons should also be made, without
skipping inconvenient cases. We know about the disappearing Aral
Sea or the pollution dangers to Lake Baikal; yet we are implicitly asked
to forget about the dried-up Baja California wetlands, the depletion
of the Ogallala Aquifer, the pollution of Lake Okeechobee, the Mississippi delta dead zone and the thinning of Lake Chad. If comparisons
are to be made, they should be made all the way. Here, I do this in a
largely illustrative manner to show some salient problems that should
be acknowledged openly and confronted. This is also because of the
dearth of data available that cross administrative boundaries or that
go beyond treating national boundaries as if they neatly demarcated
environmental impacts.

Much is unjustifiably assumed in mainstream scholarship and,
scratching the surface a bit, it turns out that three kinds of comparison underlie much of the argumentation on socialist states and the
environment: (1) absolutist, (2) synchronous and (3) diachronic.
In an absolutist comparison a social system's overall environmental
impact is assessed on the basis of the highest or lowest rankings. It
is characterised by recourse to superlatives (like worst or best) and
to the logical flaw of generalising from one or a handful of cases.
Another comparative framework is geographical and synchronous,
as in what happened in different countries during the same historical period. A third technique is to analyse environmental impacts
across time (diachronous) within a country, as in comparing the

before and after of a major historical change, like the abandonment of state socialism for liberal democracy or another capitalist variant of politics. Synchronous and diachronous comparisons can be useful in giving a general picture of the environmental impacts of socialist state systems. However, one must also beware that political units rarely if ever coincide with ecological ones. Diverse environments are enveloped within an administrative level (e.g. a country's borders) and a similar kind of ecosystem can cross different countries or administrative units. What is more, environmental impacts in the past affect the kind of environment subsequently encountered and impacted. For example, the deforestation of a landscape brings about a changed ecosystem inherited by the following generation, so that environmental impacts over the same place or region may not be comparable across time without historical analysis. Some impacts also exceed the area and period in which they occur, such as with airborne lead pollution or greenhouse gas emissions. Countries where a long-term environmental problem is detected, such as lake acidification, may not be the only or even primary source of the problem, which could be due to atmospheric transfers of pollutants from another country. These are some of the issues that make environmental impact comparisons across and within countries challenging and at times of dubious legitimacy. Changes in ecological dynamics matter as much as social ones. The utmost is tried here to account for both kinds of processes, but I also show that even comparisons that are done conventionally do not demonstrate any greater environmental friendliness for liberal democracies. In fact, the opposite is often the case.

ABSOLUTIST COMPARISONS

Using superlatives is more a propagandistic or sloganeering tactic than a methodology, but it is common enough that it merits at least a brief dissection, with an equally quick dismissal. In the conventional environmentalist imaginary, the USSR and 'eastern Europe' figure as regions of the world where the worst industrial accidents or worst cases of pollution have happened. Characteristic of this kind of ranking is an allusion to the Chernobyl accident, the destruction of the Aral Sea and so on. The argument, simply stated, is that the worst environmental insults are in state-socialist countries; therefore, the whole of socialism or communism is awful for the environment.

This 'selective citation of environmental atrocities' (McIntyre and Thornton 1978, 175) has been around for many decades. It is specious argumentation amounting to a diversionary tactic to dissimulate the main causes. A handful of industrialised or industrialising countries are claimed to represent not only all state-socialist countries, but all of socialism or all of communism, even communist projects brutally suppressed in state-socialist countries. Most state-socialist countries, though, had no such awful environmental impacts. Rather, countries like Laos and Vietnam are reeling still from the widespread and enduring environmental devastation inflicted by the US government's military invasion and carpet bombings in the 1960s and 1970s.

Another major flaw is about the lack of criteria discussed to pass judgement in the first place. Given the examples frequently cited, it can be deduced that the criteria consist of the size of the area affected, lasting negative impacts and sometimes the number of human deaths. The problem is that there are many kinds of impacts and they are not necessarily commensurable. For instance, it is true that the worst nuclear power accident occurred at Chernobyl, but it is also true that the worst oil spill in history so far happened just off the coast of Louisiana (US) in 2010, the Deepwater Horizon explosion. On land, the 1994 Mingbulak oil spill in the Ferghana Valley (Uzbekistan) is the largest recorded so far (Sharma et al. 2020), though not in terms of area and numbers of people deleteriously affected. For that, one should consider at least among the worst cases the Texaco oil drilling pollution in the lands of the Kichwa and Shuar Nations (north-east Ecuador). From 1964 to 1990, that US firm intentionally dumped 68 billion litres of toxic wastewater and 64 million litres of crude oil over a 4,400 km² area. Cancer-related deaths and miscarriages have skyrocketed among the more than 30,000 people inhabiting the area (Coronel Vargas et al. 2020; San Sebastián et al. 2001). The worst case of dioxin contamination was the 1976 industrial accident at Seveso, Italy, in terms of adverse health effects, and, in terms of permanently displaced inhabitants, the 1983 Times Beach chemical plant leak (Missouri, US). The Centralia underground coal mine fire at Centralia, Pennsylvania, has been burning since 1962, making that part of the state uninhabitable. It is unclear which of these different kinds of impacts should be considered worse than another and on what basis.

Even if the criteria were clear and took care of impact specificity, the claim about state-socialist impacts as the worst quickly evaporates. The reader may want to consider, in terms of total areas affected and loss of ecosystems, the massive losses of forests and wetlands worldwide prior to the 1917 Russian Revolution and still ongoing even after the disappearance of nearly all socialist states. In terms of lasting damage, one might wish to become acquainted or remind oneself of the Hanford radioactivity experiment on inhabitants of Washington State (US) and the nuclear detonations on Maralinga land (Australia), a case among many, within countries like the US and Australia as well as within the former USSR, where large areas remain off limits because of unsafe levels of radioactivity. It is particularly jarring to find anyone from the US, a country sprinkled with more than 1,300 highly contaminated sites (US EPA 2020a), pointing to the USSR or now China as the most polluted country. In terms of human deaths, it is under capitalist conditions where environmental disasters are accompanied by the highest mortalities. Two horrific, heart-wrenching examples are in Bhopal, where the actual worst industrial accident in history was inflicted by a US business (Mittal 2016), and the Niger Delta, where oil extraction by a combination of northern European firms and national government forces has undermined the health and livelihoods of millions.

A little past midnight on 3 December 1984, following several years of managerial negligence to reduce costs, a Union Carbide plant in Bhopal released, along other reaction products, about 40 tonnes of methyl isocyanate (used to produce carbaryl pesticide). More than half a million people were exposed to the poison, mainly poor residents, many living in shacks. Within two days 8,000 people suffocated to death, succumbing to pulmonary oedema. Over the subsequent 20 years the toll reached 20,000 deaths. Of those who survived, more than 200,000 have been harmed by life-debilitating pulmonary, gynaecological, ocular and/or neurological ailments, while thousands more were born with deformities (Broughton 2005; Dhara and Dhara 2002; Gupta and Varma 2020; Mishra et al., 2009). The plant was shut down shortly after the disaster. Years of prior wastewater discharge and post-closure leaking stockpiles have contaminated local soils and groundwater with carbaryl, hexa-chloro-cyclohexane and other persistent organochlorines, alongside chromium, mercury, nickel and lead pollution. Nearby residents, already suffering from exposure to toxic gases, remain at risk for neurological, hepatological, reproductive, endocrine and gastrointestinal damage, among other health effects, through soil contact, dust inhalation, water use and bioaccumulation (Johnson et al., 2009). The struggle for clean-up and reparations continues and those responsible for the catastrophe have yet to be brought to justice. Unlike the cases of Chernobyl and Fukushima, there were no permanent evacuations in Bhopal. People there still live in a highly contaminated environment, with little institutional support or compensation. This is in India, the world's largest liberal democracy.

In Nigeria, Africa's largest liberal democracy, there exists another under-recognised case of extensive and lasting pollution. The Niger Delta is among the world's largest coastal wetland and mangrove swamp areas (c.26,000 km²). It is home to millions of people whose livelihoods depend primarily on subsistence farming, gathering and hunting (Kalu and Stewart 2007). Since the discoveries of oil and gas reserves in the early 1950s, soils have been contaminated through onshore drilling, piping, transport and refining. Continuous gas flaring and thousands of spills are major sources of surface soil contamination, including polycyclic aromatic hydrocarbons (PAHs) and heavy metals to levels tens of thousands of times those recommended by the WHO (Abii and Nwosu 2009; Ipingbemi 2009; Iwegbue et al. 2009). Among the immediate health consequences of pollution have been deformities at birth, liver damage and dermal diseases, but carcinogenic compounds and heavy metal poisoning contribute to making many locals' lives painful and short. Soil pollution is also undermining water quality and agricultural productivity through direct crop losses and soil nutrient decline (Dung et al. 2008; Rim-Rukeh et al. 2007; Sojinu et al. 2010). Health hazards are compounded by mass impoverishment and meagre infrastructure due to the destruction of livelihood resources, heightening economic dependence on foreign institutions and the militarised appropriation by transnational corporations and the central government of the bulk of oil and gas profits. Such gross injustice, recognised in international courts, has been met since the 1970s by a combination of protests, legal action and armed struggles, without substantial redress so far (Watts 1997).

If one digs a bit more among the droves of environmental pollution cases, one discovers that the culprits of disaster are almost all private enterprises and war-related state industries from liberal democracies. The US Department of Defense is especially destructive, sending into the atmosphere more greenhouse gases than entire industrialised countries like Sweden (Crawford 2019; Schwartzman and Schwartzman 2019). But there is an even more sinister aspect to the oft-repeated proclamation on socialist states as intrinsically destructive. It is an aspect that environmental justice activists and scholars have long known about and documented (see the *Environmental Justice Atlas*;[1] Temper et al. 2015). Viewed globally, the message from Bhopal and the Niger Delta could not be louder. People of colour can be murdered at will by transnational and national capitalists without much, if any, substantive consequence. The familiar trope in the mainstream about socialism's disastrous environmental consequences conceals the fact that the worst cases have occurred in capitalist countries and that most people fatally affected by environmental destruction have not been white, much less east European whites. To borrow from Vijay Prashad (2018), the question has never really been so much about East and West as about South and North.

CATEGORISATION CRITERIA TO COMPARE ENVIRONMENTAL IMPACTS

If the claim that socialist states are the worst does not hold, perhaps one can say, as is more often said, that on the whole state socialism has been worse than capitalism on the environment. This moderated expression is a relative, rather than absolute, kind of comparison. At first glance it would seem a more plausible thesis until a comparison is attempted in more than an impressionistic or partial way, which is how comparisons have so far been done. Most of those who claim

---

1   https://ejatlas.org/.

state-socialist systems are environmentally worse furnish no criteria about which countries count as state-socialist nor justify the selection of countries included in a comparative analysis. For this and other reasons it was important to clarify these criteria in Chapter 1, so that at least those disagreeing with the outcome of the analysis offered here can spell out their own alternative criteria for comparisons and then make the comparisons accordingly (and just as systematically). What prevails instead in studies on this topic is the almost sole focus on eastern Europe, the USSR and/or China, with Cuba rarely considered, and the exclusion of state-socialist countries in Africa and South East Asia.

A systematic comparison is far from simple. There is great disparity in the total number of countries for each social system, with capitalist countries greatly outnumbering state-socialist and socialist government countries, depending on the historical period (see Chapter 1). As of 2020, the ratio of capitalist to the other two social systems combined is more than 200 to one, with Cuba argued here as the only state-socialist country left. Countries like China, Eritrea and Vietnam, ruled by (self-titled) socialist or communist single-party governments, are categorised as countries with a 'socialist government' in a capitalist economy. Social democratic parties and thereby social democracies became avowedly pro-capitalist in the aftermath of World War I. The total number of countries under different systems shifted markedly over the decades covered, as social systems were transformed from state-socialist to capitalist and some polities literally disappeared (Figure 3.1). Socialist states became capitalist regimes by the early 1990s especially in northern Eurasia and Central Asia, but also in southern and eastern Africa. Nicaragua's Sandinista government yielded to right-wingers by the early 1990s. Only a few countries, like North Korea and Cuba, did not undergo such political change, at least not in that direction, even though they were deeply affected by systemic changes in what was the USSR. As discussed earlier, in North Korea, notably, the change by the 1970s was towards the shedding of socialism in favour of Juche (self-reliance) ideology and a dynastic autocratic order. The PRC became a special case starting with a major systemic change in 1978, leading to a capitalist economy under a self-described communist party. Something similar occurred in Cambodia, Laos and Vietnam between the middle of the 1980s and the early 1990s. There are therefore multiple cut-off dates to keep in mind for different countries, although the 1989–92 period remains pivotal with respect to a capitalist turn.

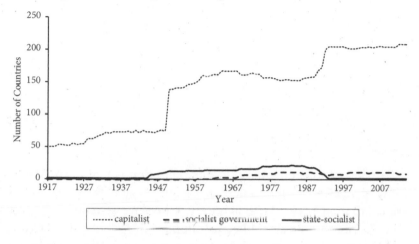

Figure 3.1   Number of countries by social system, 1960–2014

Sources: Modified from https://en.wikipedia.org/wiki/List_of_socialist states (accessed 14 January 2021); see also criteria discussed in Chapter 1.

It is striking how low the number of state-socialist countries was compared to capitalist countries even in the heyday of state socialism: the 1970s and 1980s. The low number is deceptive, though. During that time, 34 per cent of the global population lived in state-socialist countries, which covered about 30 per cent of the earth's land surface (aggregated data from CIA 1976 and UNSD 2007). About a third of the human world back then existed outside the direct rule of capitalists and certainly outside liberal democracies. Importantly, with respect to environmental impacts, the timing of industrialisation mainly coincides with the inception of state socialism or one-party socialist governments (Table 3.1; summed from Tables 1.2. and 1.3). Some countries are counted twice because they reindustrialised following war, as in Hungary, or had major manufacturing expansion on a pre-existing industrial base, as in the USSR. This is something to keep in mind because most capitalist countries continue to have low or insignificant levels of industrialisation. The potential for impact intensity is much greater in countries undergoing industrialisation, especially when, in the case of socialist states, there is no neocolonial outlet.

Table 3.1   Number of countries industrialised relative to social system introduction

Social system	Total countries	
	Preceding	*During*
State socialism	6	13
One-party socialist government	2	5

The level and timing of (re)industrialisation is tied to the availability of investments to build, maintain and expand factories, offices, transportation networks, machinery, schooling systems and much else. Economic indicators like gross domestic product (GDP) give an indication of the potential for such investments as well as for resource consumption (via purchasing power) beyond the demands of a manufacturing sector. Extreme gaps exist among countries in this respect and they cross social systems. At the same time, no country is really isolated from the capitalist world economy, as shown in Chapters 1 and 2. There are mutual influences among countries by means of, for example, commercial ties, and, fatefully for many communities, relations of domination or neocolonial interventions by the most powerful states and capitalist institutions on much of the rest of the world. Socialist states were and are enmeshed and shaped by these dynamics. A relational kind of comparative framework is therefore in order where countries' relative position in the capitalist world economy is included in the analysis. This way of proceeding with comparisons is much more apt than the typical arbitrary and highly skewed selection of only the most industrialised state-socialist countries with economic means typically reaching only half of those of the wealthiest capitalist countries against which they are compared. That is like comparing people living in luxury and privilege with those priced out of healthy foods and healthcare, and then pointing the finger at the latter for their worse health conditions or poor choices in life.

Comparisons of the environmental records of countries that interact with each other should involve, at the very least, a consideration and an analysis of data that better represent a country's relative status, that imply a wider interlinked context and that also point to processes shaping all sorts of environmental policies and practices. Drawing from world systems scholarship, this can be more effectively accomplished by using the analytical category of world-system position, which is historically dynamic (the position can shift) and accounts for enormous power differences. Relations of power, in this case at the international scale, are represented by the categories of periphery, semi-periphery and core. Minqi Li masterfully and succinctly describes them this way:

> The core regions, because of their strong military power and monopoly over the leading sectors of the capitalist world economy, are able to extract economic surplus from the periphery and the semi-periphery through unequal exchange. Much of the economic surplus produced by the periphery is extracted by the core and the semi-periphery. The semi-peripheral regions extract economic surplus from the periphery but are exploited by the core. (Li 2016, 197)

Minqi Li's per capita GDP method is employed here, along with the same database (Bolt et al. 2018), to approximate world-system position and account for population size. Countries with more than double the world's average per capita GDP are assigned core status and those in the semi-periphery have values from average to twice the average figures. Below average per capita GDP places a country in the periphery. This way, the most powerful state-socialist countries like the USSR, German Democratic Republic and PRC are compared with much more like analogues in the semi-periphery of the capitalist world, like Brazil, India, Mexico and Saudi Arabia. The wealthiest capitalist countries in the world are in a sordid league of their own and should be treated as such – not as exceptions, but as countries embodying a substantively different status as the main manipulators of the world economy and enforcers of capitalism (Amin 2018; Wallerstein 1979). At the same time, and this is a major departure from standard fare, the state-socialist periphery is included in the analysis and compared to like capitalist regions characterised by similar economic conditions.

Within the fifty-year period considered (1960s through the 2010s), shifts in position have had negligible statistical effects. Nevertheless, for the countries where this applies the average ranking is computed and rounded to the nearest integer. The decades and number of countries considered vary because of data availability limitations. Most countries are represented (86–91 per cent, the higher percentage starting in the 1990s), and data for small island or archipelago countries tend to be under-reported. The historical time frame includes the maximum extent of state socialism as well as the major bouts of industrialisation state-socialist countries underwent. Thus, the very worst environmental effects of state socialism are also displayed. It turns out that even using the least flattering data for state-socialist countries points to their environmental impacts being less terrible than those of their capitalist counterparts.

COMPARING SOCIAL SYSTEMS OVER THE SAME PERIODS (SYNCHRONIC COMPARISONS)

A general comparison between social systems, while effacing differences among countries, is still useful to get a panoramic, global view of environmental impact as long as no more than that is said about such a comparison. This kind of data analysis is done on CO_2 emissions and ecological footprints of consumption because of their planetary, long-term consequences through climate change and pressures on the biosphere, respectively. Details on these parameters and terms are provided later in this chapter and standard deviations are visually displayed regarding CO_2 emissions to show how wide discrepancies among countries can be acknowledged. CO_2 is closely linked to fossil fuel combustion, the mainstay of energy use, while raw material extraction and resource use in general affects the broader impact on ecosystems. Hence, and using per capita figures, biophysical processes, economic processes and human well-being aspects are being considered at once. CH_4 (methane) emissions also have global climate repercussions, but they are entered into a comparative analysis based on both world-system position and social system type to explore a more relational angle to greenhouse gas emissions. To represent other atmospheric emissions bearing more regional or localised impacts, SO_2 emissions are analysed relative to world-system position and social system. This is because sulphur emissions typically form part of the more egregious insults stemming from the coal-based energy use typifying the state-socialist countries that engaged in rapid industrialisation. The point of all this is not to offer an exhaustive synchronic comparative study, but to show, by illustration, the baselessness of popular notions about the environmental impacts of state socialism, even according to those notions' own logic.

CARBON EMISSIONS

CO_2 comprises most of the emitted greenhouse gases linked to climate change (roughly three-quarters of all greenhouse gas emissions). When average per capita emissions of CO_2 are scrutinised (Figure 3.2a), capitalist countries in short order evince a much greater impact. Because they were rapidly industrialising or reindustrialising to overcome the devastation of war, state-socialist countries had greater impacts in the early 1960s, while most of the capitalist world remained rural and many countries were just gaining formal independence. Many of the most highly industrialised capitalist countries were also reeling from a couple of wars, so mass consumption rates were just picking up speed, and there were still parts of the economy that were not tied to fossil fuel consumption. For example, household appliances were not widespread. This is what depressed average values for capitalist countries until the mass consumption boom spread, mainly in the core capitalist world.

In the middle of the 1960s, capitalist countries, mainly the highly industrialised ones, accelerated fossil fuel use much more and simply took off with CO_2 emissions, dwarfing the state-socialist and socialist governed countries. Non- or anti-Marxist, self-described socialist, one-party ruled countries mainly came into being between the late

(a)

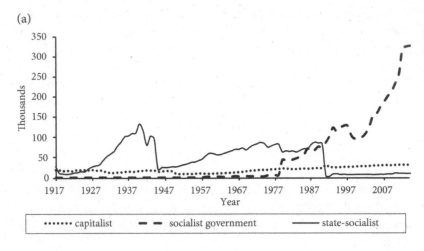

(b)

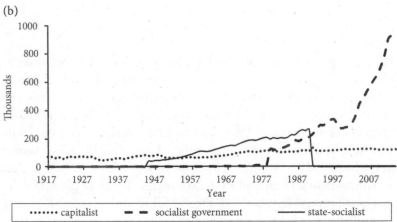

Figure 3.2 (a) average annual per capita CO_2 emissions (tonnes) per country by social system, 1960–2014; (b) standard deviation of annual per capita CO_2 emissions (tonnes) per country by social system, 1960–2014

Sources: (a) annual CO_2 emissions data are from Crippa et al. (2019) and grouped and averaged according to social system for countries in continuous existence. Per capita CO_2 emissions for the German Democratic Republic, the USSR and Yugoslavia are from Boden et al. (2011); (b) Annual CO_2 emissions data are from Crippa et al. (2019) and converted to per capita figures using demographic data from the World Bank (2019) for countries in continuous existence. Per capita CO_2 emissions for the German Democratic Republic, the USSR and Yugoslavia are from Boden et al. (2011).

1950s and the late 1960s, with countries like Iraq and Libya accounting for the spike in CO_2 emissions as a result of increasing fossil fuel production capacity and to some extent industrialising. In state-socialist countries, emissions in the late 1960s and through the 1970s were tempered by the entry of agrarian or largely peasant societies becoming state-socialist, such as Congo, Laos, Kampuchea and Yemen. State-socialist countries were mainly composed of countries with low to middle GDP.

In the case of state-socialist countries, there is another factor seldom considered except when criticising socialism. GDP is, strictly speaking, based on market valuation, hardly the mainstay of state-socialist economic policies. Because market-based valuation and profit rates, which means constantly raising consumption rates, are typically shunned, if not considered illegitimate, in state-socialist countries, GDP or similar economic indicators may be less appropriate than total volumes of resource inputs and use-value outputs. Nevertheless, GDP is highly correlated with energy consumption, which is mostly derived by burning fossil fuels, which emit greenhouse gases (Schwartzman and Schwartzman 2019; Tucker 1995). State-socialist per capita emissions plummeted by the early 1990s, as most were transmogrified into capitalist systems or became single-party socialist government countries with capitalist economies, like the PRC and Vietnam. These latter countries intensified industrialisation and became integrated into the world economy as manufacturing centres, but it was primarily in China where the bulk of per capita yearly emissions occured (see Chapter 4). A steady increase occured through the 2000s throughout the capitalist world, mainly with the expansion of manufacturing in countries with a middle GDP level, including India and Brazil. State-socialist averages only reflect Cuba's figures after 1992, and this points to Cuba being below average by two or more orders of magnitude relative to capitalist countries. This is the probable reason for the tendency of countries with lower or decreasing GDP over time contributing less greenhouse gas emissions.

Averages mask enormous differences among countries within each system, such as the much greater emissions in places like the US and the Persian Gulf monarchies. These differences can be recognised by including standard deviation values (Figure 3.2b). Doing so also makes evident Cuba as the sole remaining state-socialist country since 1992 (hence the line interruption). The standard deviation for state-socialist countries is much more contained compared to capitalist and socialist government counterparts. The observable spike at the end of the 1960s is traceable to major oil exporters becoming nominally socialist in government, like Iraq and Libya, but consumption levels rapidly decrease, and it was mainly with the transformation of the PRC economy since 1978 that per capita emissions started rising again. However, the standard deviation is not as high as for capitalist countries, even if such figures include countries with the lowest GDP values in the world. Standard deviations over time indicate large variability among capitalist countries until the formerly state-socialist industrialised countries, especially in central and eastern Europe, joined the capitalist fold. Countries with socialist governments display large differences only when fossil fuel-producing countries started raising output, but afterwards, with the expansion in the number of such countries, the levels of per capita emissions by country first dropped as a result of non-industrialised countries and then jumped by the early 1990s with mainly the PRC contributing. In part this is also the consequence of the PRC becoming a net oil importer by 1992. Standard deviations for other greenhouse gases and air pollutants follow the same patterns as in the Figure 3.2 and for the same reasons, so they are not reported or discussed further. Noteworthy is that conversion into capitalist systems does not lead to an overall decline in per capita emissions. On the contrary, after the transition to capitalism, per capita emissions increase, while generally the opposite happens with conversion to state socialism. Though increasing especially since the late 1950s, world CO_2 emissions ballooned from the early 1990s onwards, in the absence of any substantive influence by socialist states.

Emitted CO_2 lingers in the atmosphere for decades to more than a century, bearing lasting effects. So, it is important to consider cumulative (historical) contributions to gauge the impacts of different social systems, not just yearly emissions. The maximum atmospheric residence time of CO_2 as well as available data coincide approximately with the time that has elapsed since the Russian Revolution. This makes for a direct historical comparison, except that both prior to 1946 and after 1992 only one state-socialist country existed. A more appropriate comparison would then be for the 1946–92 period, at the height of global state-socialist presence encompassing roughly a third of humanity. During that time about 69 per cent of the world cumulative total belonged to the capitalist countries and one could add the 5 per cent from the countries under one-party socialist governments, since they presided over capitalist economies (Table 3.2). Historically, then, capitalist economies have disproportionate carbon emissions, exceeding the two-thirds of the world they occupied in the 1946–92 bracket. This disproportionality argument is only strengthened when including capitalist markets under one-party socialist governments. Not only that, but there is an enormous emissions differential between the time of the maximum number of socialist states and when one socialist state existed. This suggests, as others have noted (Baer 2018, 32), that with socialist states in retreat much more CO_2 is emitted globally, more than twice the amount (1,518,826 compared to 624,081 million tonnes). It is as if socialist states substantially restrained capitalist excesses. Perhaps a renewed worldwide spread of state socialism could more than halve capitalist countries' emissions, but this is sheer conjecture. On the other hand, if the

Table 3.2 Cumulative CO_2 emissions aggregated by social system (million tonnes)

	World	Capitalist	One-party socialist government	State-socialist
1946–92				
Total	624,081	431,270	31,487	161,324
(%)	100	69	5	26
1917–2018				
Total	1,518,826	1,150,617	20,619	162,014
(%)	100	76	13	11

Differential (pre-1946 and post-1992; or, when state socialism exists in only one country)

Total	894,745	719,347	174,708	690
(%)	100	80.4	19.5	0.1

Source: Ritchie and Roser (2017); country-level data are aggregated according to social system category criteria (see Chapter 1).

Cuban model were replicated, precipitous reductions and overall improvement globally would be assured.

METHANE AND SULPHUR EMISSIONS AND POSITION IN THE CAPITALIST WORLD ECONOMY

Methane (CH_4) is a greenhouse gas with greater global warming potential than CO_2. Once airborne it can stay in the atmosphere for more than a decade and is 50 times more powerful than CO_2 in terms of global warming effects. Thankfully all countries show a precipitous decline in CH_4 emissions since 1970 in the core and semi-periphery (Figure 3.3). The trend for the industrialising or industrialised state-socialist countries in the semi-periphery, however, is one of very low emissions throughout, even as they increasingly shifted to methane to counter pollution from coal combustion. Most notably, core capitalist countries emit the most per capita on average. This is almost the reverse for the periphery, where socialist states like the PRC were continuing with rapid industrialisation during that period. However, the emissions, even at their peak, reach but a small fraction of cap-

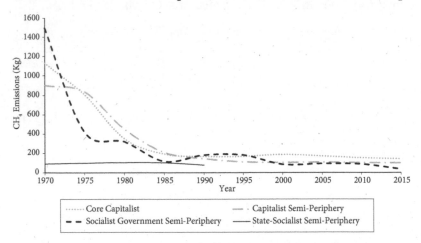

Figure 3.3   Per capita CH_4 emissions (kg) by social system relative to core and semi-periphery position in the capitalist world economy, 1970–2015

Sources: CH_4 emissions data are from EC-JRC (2019). Per capita GDP figures are from Bolt et al. (2018), standardised as constant 1990 international dollar values. Those data are used to determine world-system position as described in the text.

italist core and semi-periphery countries. There was a very quick reduction by the 1980s, reflecting the systemic change in China and the mitigating effects of low-emission, lower-input agrarian countries like Angola, Madagascar and Mozambique into the state-socialist camp by the middle of the 1970s. The sudden reduction reached a new emissions rate that remained stable throughout the 1980s and then dropped even further, to below levels in both the capitalist and one-party socialist government periphery (Figure 3.4). Overall, relative to CO_2 and CH_4 emissions, capitalist systems consistently impact the atmosphere much more than either state-socialist or one-party socialist systems.

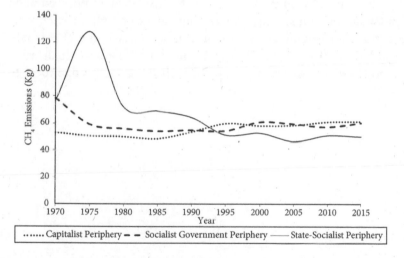

Figure 3.4   Per capita CH_4 emissions (kg) by social system relative to periphery position in the capitalist world economy, 1970–2015

Sources: CH_4 emissions data are from EC-JRC (2019). Per capita GDP data are from Bolt et al. (2018), standardised as constant 1990 international dollar values. Those data are used to determine world-system position as described in the text.

Comparisons, as stated above, should be thorough, and so presented here is also evidence of a net negative state-socialist impact on the atmosphere. Air pollutants like sulphur emissions (SO_2), also a main ingredient of acid rain (Grennfelt et al. 2020), reflects less favourably on socialist states, and nowadays even more so with one-party socialist states, mainly the PRC (Figure 3.5). There is an irony to this. As explained in Chapter 4, the USSR government's successful efforts to set up an international long-range air pollutants reduction agreement is a major factor in the SO_2 emissions decline in core countries from the late 1970s onwards. Aside from that, this rare, noteworthy accomplishment in core countries is due to sustained environmentalist struggles forcing heavy government regulation on polluting capitalist firms. Data comparisons of this sort conceal the existence of such struggles. It should also be kept in mind that the lower levels in core countries are still four times those of the one-party socialist government and capitalist peripheries.

Another matter needs to be clarified as well. The high level of sulphur emissions is largely because of a historically unprecedented rapid pace of industrialisation relying on often sulphur-rich coal reserves, more abundantly available in countries like the USSR, Poland, the PRC and the German Democratic Republic. Coal-burning power plants were not equipped with the cutting-edge tools to

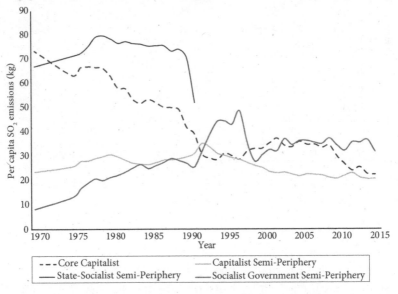

Figure 3.5 Per capita SO_2 emissions (kg) by social system relative to core and semi-periphery position in the capitalist world economy, 1970–2015

Sources: SO_2 emissions data are from EC-JRC (2020). Per capita GDP figures are from Bolt et al. (2018), standardised as constant 1990 international dollar values. Those data are used to determine world-system position as described in the text.

reduce sulphur output, as in core countries. At the same time there were measures taken by the 1980s in industrialised state-socialist countries to attenuate emissions and this is shown by a steady decline in SO_2 emitted (Krüger et al. 2004; Welford 1991). The pattern is therefore similar historically to what transpired in the capitalist core (Dominick 1998), but this observation is to some extent deceptive. Because the peak pollution rates occurred over a much more compact period, the negative impact was also shorter in duration. Had socialist states continued their path of amelioration, SO_2 emissions would have continued declining.

One-party socialist systems with capitalist economies exhibit a major sulphur emission uptick in the 1990s, mostly explainable by the PRC's entry into the fold. The massive increase in emissions meets the rates featured in the disappearing state-socialist systems of the semi-periphery. There have been effective attenuating and preventive measures that have succeeded in cutting emissions down by at least a third, but the rates remain the highest. Alongside major industrial output increases, this is also traceable to the privatisation reforms and rapid industrialisation in the PRC combined with the influx of direct investment from capitalist countries, shown often to correlate highly with increases in, for example, CO_2 emissions (Sarkodie et al. 2020). This is taken up again in Chapter 4. Regardless, with the capitalist transformation of nearly all socialist state countries, whether inclusive or not of formal political institutions, the net effect now is that half of humanity, mainly in India and China, is exposed to rising air pollution, resulting in millions of premature deaths every year (Shaddick et al. 2020).

The capitalist periphery had a consistently lower impact relative to SO_2 emissions compared to the socialist-oriented camps until recently with sulphur emissions (Figure 3.6). However, the post-1990 figures exhibit a major jump in the state-socialist category. This is due to Cuba being the sole state-socialist country remaining by then, so the trend from the 1990s refers to that of Cuba alone. Relative to most of the periphery, though, Cuba probably ranks highest in levels of industrialisation, urbanisation and overall standard of living, so the comparison gets tenuous. After a sudden fall, emissions reached levels comparable to those in the semi-periphery (with emissions in the range of Taiwan's and the PRC's, for example), which makes sense in terms of raising or maintaining high human development levels. This occurred as the economic output recovered from the sharp downturn linked to the disappearance of the USSR. In part, the problem is due to the US embargo, which impedes the local development and/or transfers of technologies from abroad that would help prevent the release of SO_2 into the atmosphere. As discussed in Chapter 4, there are major positive changes in Cuba that are leading to reversing this trend.

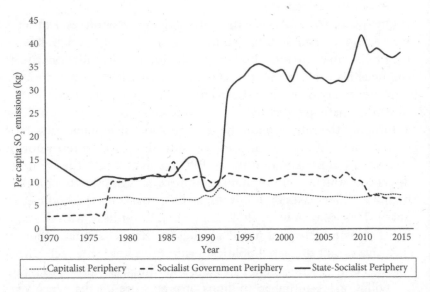

Figure 3.6   Per capita SO2 emissions (kg) by social system relative to periphery position in the capitalist world economy, 1970–2015

Sources: SO2 emissions data are from EC-JRC (2020). Per capita GDP data are from Bolt et al. (2018), standardised as constant 1990 international dollar values. Those data are used to determine world-system position as described in the text.

OTHER BIOPHYSICAL PARAMETERS

So far, taking a partially conventional route to comparisons and focusing on the atmosphere, the picture is an equivocal one. It does not support the contention that state socialism is worse on the environment. Arguably, impacts on the chemical composition of the atmosphere are easier to compare, too. The atmosphere has less geo-graphically and historically variable characteristics and processes than ecosystems, soils and surface waters on continental landmasses. Air emissions data are also more complete than most other kinds of environmental data. Effects of pollutants discharged into oceans and seas are tougher to analyse because the pollution sources are much more difficult to pin down to individual countries and the available data are not as geographically encompassing. Studying the spread of coastal dead zones is also at times challenging in terms of account-ing for the wide variability of coastal ecosystems (from wetlands to semi-deserts) and, to some extent, in terms of determining prove-nance, when rivers cross different countries. Comparing impacts like deforestation, groundwater withdrawal, or soil erosion requires accounting for the great variety of forest ecosystems (with different biodiversity potentials), aquifer types (with differing capacities and replenishment rates) and soil types and their topographical position, while also controlling for the effects of past land uses under different social systems.

Moreover, data are often insufficient or inadequate and monitor-ing systems may be spotty. Monitoring systems are distributed in extremely uneven ways worldwide (on soils data and monitoring, see Engel-Di Mauro 2014). In most countries, governments have limited infrastructure and budgets at their disposal even to carry out a mon-itoring programme. The matter cannot be easily resolved by using remote sensing data because global coverage is relatively recent and sometimes not as reliable as direct monitoring techniques. A more ecologically attuned alternative would be longitudinal (diachronic) studies of the same kinds of ecosystems within the same countries or, even better, when studies are available, ecoregions across countries. This is taken up below.

Regardless, pretending all the above problems away does not yield the sort of results that would support the contention that state socialism is intrinsically damaging. In the case of soil degradation, this is rather evident. Food and Agriculture Organization (FAO) data compiled in the late 1980s show the USSR as no worse than other countries. In fact, the ranking of the USSR was 28th, much higher than that of the US, which was 52nd (higher ranking meaning less overall degradation). If one extends the comparisons, one finds Hungary ranked similarly higher than the Netherlands and other such results that indicate less soil degradation in industrialised state-socialist countries relative to the industrialised liberal democratic counterparts. Other interesting comparisons can be made, such as the Koreas having virtually the same ranking and Vietnam and Laos having much better rankings than Thailand, which is in turn ranked lower than Cambodia (Bot et al. 2000).

These results are especially interesting in view of the exaggerated statements made by pro-capitalist institutions and intellectuals. For example, based on a 1988 study, the IMF, World Bank, Organisation for Economic Co-operation and Development and European Bank for Reconstruction and Development posited that many regions in the former USSR were on the verge of ecological breakdown (reported in Hill 1997, 1). If that were true, the US should have already reached soil degradation hell by the 1980s, given the results of the above-discussed FAO study. Things get barmier still. Relying on just one 1995 World Bank report by a Hungarian agricultural economist, Josephson et al. (2013, 212) assert that a quarter of farmland in Armenia was eroded away 'as a result of intensive and irrational agricultural practices' during the USSR period. This is quite a statement, considering that in Armenia the only output that did not decline in the 1990s was in the farming sector, according to other agricultural economists (Lerman and Mirzakharian 2001, 9). Similar kinds of exaggerations, through the same World Bank report, are spread about Azerbaijan and Belarus (more than 20 per cent arable land loss) and Ukraine (a third of farmland lost). Strangely enough, people in those countries are not experiencing permanent food production shortfalls or mass starvation.

As intimated, these kinds of soil degradation studies must be interpreted with much care, as they lump together information that is not easily comparable. For soil erosion severity comparisons one must control for more and less erodible kinds of soils. Countries with much steep topography, earthquake potential and/or active volcanoes should not be compared with countries mostly made up of plains in seismically inactive regions, at least not without qualification and apposite parametric adjustments. There are also problems related to the criteria used. Most authors of soil degradation studies measure potential land degradation relative to the requirements of an industrialised, market-oriented farming system, rather than small-scale peasant farming needs or hunter-gatherer requirements. Nevertheless, the much more credible FAO study, carried out by soil scientists rather than economists, directly contradicts the claims about state-socialist systems having somehow greater destructive propensities, even when using the same terms and the same biases as those implicit in those claims (e.g. assuming land to be arable if it yields marketable produce).

ECOLOGICAL FOOTPRINTS

There is another way to compare overall impacts that in some way bypasses the above-described challenges. The method is called the 'ecological footprint'. It is an estimate of how much of the biosphere's regenerative capacity is taken up by human activities, measured in global hectares (Moran et al. 2008). A global hectare (gha) is an area with a biological regeneration rate equal to that of a world-average biologically productive hectare (every gha has the same amount of bioproductivity). This method still glosses over wide differences, if not incommensurables between ecosystems and geospheres, but it is a more accurate parameter than soil degradation, water withdrawals, deforestation rates and such. This is because the distribution of soil types, forests and water resources vary independently of social system. For instance, it is entirely inappropriate to compare deforestation rates between a country endowed with lush rainforest as the default ecosystem with a country that has little to no forests even without any human-caused deforestation. Likewise, it would be just as problematic to compare soil erosion rates in a highly mountainous country with those in a country wholly located on a plain and accruing sediment through periodical floods (fresh material for soil formation) from upstream sources located in neighbouring countries.

In part to circumvent biophysical incommensurability, the analysis here relies on the Ecological Footprint of Consumption (EFC). The EFC is computed by adding the net value of ecological footprints from imports and exports (in gha) to the sum of all products from different uses on land and sea. This estimate provides a major advantage over the comparisons done above because imports and exports are included. In this manner, EFC comparisons enable some accounting for global interconnections, which are typically left out of the picture. With these qualifications the per capita EFC values aggregated by a social system reveals patterns not too dissimilar from the ones described above, with capitalist social systems disproportionately the culprit of planetary destruction through excessive resource consumption (Figure 3.7).

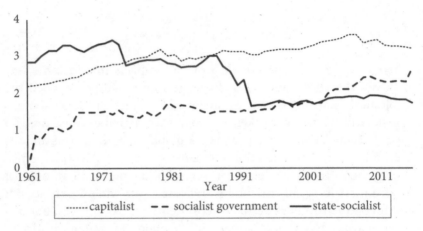

Figure 3.7   Average per capita EFC per country by social system (global gigahectares), 1961–2016

Source: Global Footprint Network (2019).

The 1960s were dominated by the rapid industrialisation drives of some of the state-socialist countries aligned with the USSR or the PRC. Through the 1980s, Czechoslovakia, Poland and the USSR reached the second decile. The result is a greater impact than capitalist countries, which comprise a great number of countries that were starting to gain formal independence and with low to negligible manufacturing output and largely exporting raw materials. But by the middle of the 1970s, average EFCs in state-socialist countries started declining to levels below those of capitalist countries. This should be expected, as countries like Ethiopia, Madagascar and

Mozambique started joining the state-socialist camp. It also needs to be underlined again that most capitalist countries were not industrialised and served mainly as suppliers of raw materials (and still do). This explains why average values for greenhouse gases, pollutants and the EFC tended to be higher for state-socialist countries. State-socialist countries did not have former colonies to exploit through neocolonial relationships and this meant a far greater level of within-country pollution from the combination of raw material extraction and industrial output. The fact that, by the 1980s, the average per capita EFCs reached levels similar to or lower than those of capitalist countries should be regarded as a major feat. This trend continued until the sharp drop of the 1990s, as the number of state-socialist countries dwindled to one, Cuba, where EFC levels by the 2000s were markedly lower than all averages.

Countries under single-party socialist governments (e.g. Libya, Syria) nationalised and developed some industries, but were largely exporters of raw materials as well. Nevertheless, their per capita EFCs have been much lower than the capitalist average from the 1960s to the present. Importantly, such countries include the PRC after 1978 and Laos and Vietnam after 1986. Per capita EFCs still managed to stay substantially lower than those of capitalist countries, irrespective of the massive growth in consumption levels and high integration into global trade and investment flows. In fact, the per capita EFCs of capitalist countries kept growing over time, barring the decrease and stagnation of the recession-fraught 1980s, and somewhat mirrored in state-socialist countries (belying notions of state-socialist economic insulation). The recent slight drop coincides with the 2008 financial collapse, which had little effect on resource consumption rates in the PRC, among other less impacting countries. The gap may be closing between socialist government and capitalist systems, but – at least relative to total resource consumption – it cannot be said that the PRC is the most negatively impacting country ecologically. In fact, according to 2016 data China ranks 68th out of 185 countries in per capita EFC. The situation is even starker when total EFCs are divided according to social system (Figure 3.7). Capitalist countries are simply inordinately destructive. That average per capita EFCs continue to grow in the 2000s just demonstrates how awful to the planet capitalism is, regardless of its political variant.

The net effects of EFCs can be compared to the biocapacity within a country's territory to determine whether the country is in ecological deficit or surplus (reserve). When a country's EFC exceeds its biocapacity, the country is deemed to be in ecological deficit; the converse is to have an ecological reserve. Biocapacity is the biosphere's productive capacity and socially useful resource provision. Given the highly uneven distribution of different kinds of ecosystems and physical environments, treating all countries as if they could rely on their own biocapacities within the same global system (capitalism) skews results in favour of the most ecologically and environmentally endowed countries, especially those encompassing multiple biomes. This is one reason for conducting more contextualised, country-specific studies, as in Chapter 4.

The issue of resource usefulness could also pose some difficulty because industrialised capitalist systems are taken as the norm regarding what is useful and, moreover, what counts as resource changes over time. For example, rare earths are now much more sought after than in the past, thanks to products like mobile phones and computers. However, national-level data already preclude consideration of differing social systems and their differing resource bases at subnational scales (e.g. Kanien'kéha communities in Canada and the US). The fact that socialist states and one-party socialist governments concentrated on industrialisation and/or raising output – even when for different ends – makes such countries more comparable to capitalist systems, even if they are politically light years away from each other.

With these qualifications in mind, capitalist countries overall show a precipitous decline in ecological sustainability from a large ecological reserve to a net deficit by the end of the 1970s (Figure 3.8). State-socialist countries were only slightly in reserve and quickly got into deficit as industrialisation was achieved in some of the countries and large resource extraction projects were introduced in countries with largely peasant economies. By the middle of the 1970s and through the 1980s, both systems were in ecological deficit, alternating with each other in terms of degree. On average, capitalist countries were less ecologically demanding. This reflects the much larger average area that capitalist countries occupied, which is almost the entire tropics, where the greatest biocapacities are located relative to ecosystem biomass productivity and biodiversity. However, by the end of the 1980s the state-socialist systems' deficit was being markedly reduced just as they were being made to disappear. Their average was converging with that of capitalist countries. Cuba constituted the only state-socialist country by the 1990s and the low deficit testifies to their having become the most biophysically sustainable country relative to the standard of living. Socialist government countries initially hovered close to parity until the PRC joined, which steadily dragged the impact into increasingly larger deficit. However, on aggregate such countries remained in less deficit than capitalist ones.

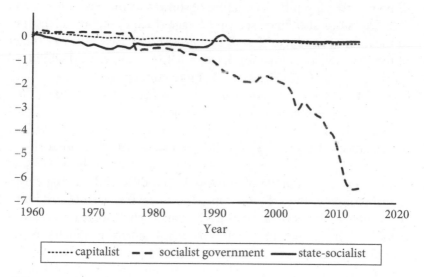

Figure 3.8 Net ecological reserves or deficit by social system (global gigahectares), 1961–2014

Source: Global Footprint Network (2019).

COMPARING BEFORE AND AFTER SYSTEMIC CHANGE (DIACHRONIC COMPARISONS)

A third way of making comparisons is by looking at environmental impacts longitudinally, that is, before and after major social transformations. By the early 1990s most state-socialist countries became capitalist, so just reinspecting the above-displayed graphs can already give a sense of the environmental effects of expanding the capitalist world. From the above analyses it is evident that the systemic changes in formerly state-socialist countries have hardly brought any improvement on previous environmental impacts. Total air pollutant and greenhouse gas emissions have increased sharply since the 1990s, including in the new liberal democratic variants of central and eastern Europe, swallowed up first by NATO and then by the European Union. In part this is due to the progressive disinvestment in public transport and the spread of individual motorised vehicle use. Average emissions and ecological footprints have either remained roughly as high as before or somewhat increased. Per capita CO_2 emissions also increased on average since 1990, but at least not to the level of the 1970s. Recall that greenhouse gas emissions linger in the atmosphere at least for decades before they are washed out, so a constant rate of emissions means more cumulative damage.

A study in *Nature Climate Change* from Peters et al. (2012) on the relationship between economic crises and CO_2 emissions confirms this disastrous trend but adds even more revealing connections. The demise of most socialist states did not lead to as sharp a decline in CO_2 emissions as the US savings and loans scam of the 1980s and the US real estate bubble of 2008–9. Stated differently, a single capitalist economy, the US economy, has been responsible for much more intense climate change impacts than all state-socialist countries combined. It is therefore not only the historically much greater cumulatively destructive impact of industrialised liberal democracies that is at issue, but also the intensity and weight of the annual, shorter-term impact, which becomes more evident by comparing yearly economic growth and CO_2 emissions figures.

SO_2 emissions are the exception. They have remained roughly at the same level of destructiveness over time. For this less terrible outcome one can probably thank the restraining influence of transboundary pollution treaties, where the USSR played an essential role as initiator (see the first section in Chapter 4, 'The USSR: Creating Mass Ecological Consciousness'). Whether this is a great stride within a capitalist world economy is disputable. Globally, much of the sulphur emissions has been shifted to the PRC, largely through coal burning to feed into manufacturing products mainly destined for the wealthiest capitalist markets, like the US. Perhaps this is what made possible the successful implementation of transboundary pollution treaties in Europe and North America by the 1990s, the sort that some Westerners praise and indicate as an example to follow (Maas et al. 2016). If the thesis presented here is correct about the PRC acting as the pollution redirection valve, one should find air pollution treaties of that ilk rather disconcerting.

The form and amounts of air pollution have shifted substantially within former state-socialist countries. In the days of the USSR air pollution reached a peak of 75 per cent of what was emitted in the contemporary US and was declining, thanks to an increasing switch to natural gas and to successes from environmental policy application. The systemic change brought economic collapse and so pollutants from stationary sources decreased even further. They had picked up again by the 2000s, with increasing output and less pollution oversight. With the expansion of motorised vehicle use (the main mobile source), nitrous oxide and carbon monoxide emissions rapidly increased by about 20 per cent within a decade (Oldfield 2005, 100). Something similar has occurred in countries like Hungary with the production and disposal of waste. It has risen markedly and in conformity with contradictory EU policies that ultimately encourage the multiplication of waste (Gille 2004).

The net effect of systemic change on forest ecosystems is mixed and tending to the negative. This is part of wider global trends. Since 1990 the increasing deforestation rates and total losses in forested areas have been staggering (Houghton 2016; Williams 2006). About 178 million hectares of forest have been chopped, equivalent to a bit less than the area of Indonesia or Mexico. Encroachment into tropical forests have also led to a rise in zoonotic diseases (WWF 2020, 16). What is more, since 1991 every 1 per cent wealth increase from cash crops investments by a handful of capitalists resulted in a 2.4 to 10 per cent cash crop area enlargement at the expense of forests in Latin America and Southeast Asia (Ceddia 2020). These linkages and wider lenses are ultimately decisive when assessing longitudinal changes, not only in forest cover within state-socialist countries but also in ecosystems more broadly. Viewed globally, the impacts of socialist states are mild in comparison to the global capital onslaught on the biosphere.

Nevertheless, forest ecosystems have a mixed relationship with state-socialist systems, with some periods of much afforestation (in part overturning pre-revolutionary deforestation legacies in countries like the USSR and China) and others of high regional losses due to logging or pollution. This mixed outcome nevertheless had a net positive in forest expansion. Systemic changes in the late 1980s and early 1990s have not improved matters or have led to worse results (for countries like China, the early 1980s is a more appropriate chronology). Here, examples are drawn from existing longitudinal studies mainly from central and eastern Europe. The outcome for forest ecosystems of a turn to liberal democracy tends not to be terribly positive.

In the western Caucasus region, annually logging rates between 1985 and 2010 were low (0.03 per cent). Instead, the biggest clear-cutting episode happened in preparation for the 2014 Winter Olympics (Bragina et al. 2015a). The dismemberment of Yugoslavia, where private smallholding was not uncommon, did not necessarily lead to environmentally sustainable forestry. In northern Croatia, privately owned woods were sites of more destructive logging and have remained so since the 1990s, with a decline of 1.8 per cent. However, in the state-owned or managed forests the woods have expanded by 10 per cent. This is without considering the issue of forest fragmentation due to parcelling out of land through 'restitution' or sale. Regardless, the extent of privately owned forest has shrunk from 24 to 19 per cent since the violent destruction of state-socialist Yugoslavia. It is, in this case, an expansion of state ownership, not privatisation, that has enabled the preservation and even increase of forests (Cvitanović et al. 2016).

One interesting set of studies comes from the Carpathian ecoregion. It is an area of about 380,000 km² or roughly 6 per cent of the part of central and eastern Europe that used to be state-socialist. The studies show a less than stellar situation following 1990. Under Polish administrative boundaries, the Carpathian forest witnessed a sharp, rapid fragmentation and contraction between 1988 and 2000, whether the woods were owned by private, public or national park entities. Privately owned forests were chopped five times as much. It must be added that in Poland farming was mainly in private hands under state socialism (Kuemmerle et al. 2009). So whatever negative impact came about through farming in Poland is hardly due to 'socialism'. Conversely, forest regeneration since 1990 happened as well because of farm abandonment.

In the Carpathian parts of southern Romania, as much as a fifth of farmland was abandoned after 1990 (Kuemmerle et al. 2009), compared to roughly 24 per cent for the entire Carpathian region. But by 2010, as with most of the Carpathian region, the more profitable lands came to be recultivated, especially with EU membership, with the much smaller remaining land area turning into forest or grassland. Overall the initial sizeable abandonment of cropland in the Carpathians ecoregion did not end up churning out much reforestation or grassland expansion except in areas now at the neocapitalist margins, like the eastern Hungarian and Ukrainian portions of the region (Griffiths 2013). Over a much longer term, the last 250 years, most deforestation happened prior to socialist state formation but what remained of forested areas were also relatively stable. Forest areas also tended to remain stable or expand during or close to calamities, like the world wars, systemic change in the 1990s and EU accession in the early 2000s. This is excepting the Romanian portion of the Carpathians, where net loss has occurred since 2000. However, even more interestingly, most forests stayed or were enlarged during socialist state rule, especially on the northern slopes (Munteanu et al. 2014). Results from a similar study for parts of rural Slovakia show similar patterns (Kanianska et al. 2014).

Curiously enough, some of the same Carpathian ecoregion forest researchers claim a net reforestation tendency for all of central and eastern Europe. This is allegedly linked to an uneven disappearance and rebounding of farming. But such a view is rather suspect because it is generalised from just a single county or a single national park within Albania, Latvia and Romania (Taff et al. 2009). For Albania and Romania in particular, it seems the researchers lost the forest for the trees, given how well documented the post-1990 logging frenzy has been for those two countries, as reported by the very same authors (Taff et al. 2009, 5). Other studies, using satellite rather than officially reported data, show Latvia as a site of net forest loss during 1985–2012 (Alix-Garcia et al. 2016, 287). At the scale of the Carpathian ecoregion, a fifth of forestland has seen the replacement of entire stands since the late 1980s. The slight forest recovery in the ecoregion since the systemic change only in part countered widespread clear-cutting and other less intense logging. What is more, the worst period for Carpathian forests was from the late 1980s through the early 1990s and since 2000 (Griffiths et al. 2014).

As the last cited study indicates, there are important issues to be considered when findings point to forest rebound. First is that forest ecosystems should not be treated solely in terms of number of trees irrespective of species. It may make sense if one wants to reduce living organisms to how much carbon they can store, but this way there is no difference made between tree plantations or monocultures and biodiverse forests. Second, just because trees are returning to what was previously farmland, there is no guarantee that the resulting woods will be contributing to more biodiversity compared to the agroecosystems preceding them, that the overall outcome helps reduce soil erosion or even that they will become net carbon stores. In fact, there is no guarantee that woodland will replace abandoned cultivated land at all. In a study carried out in Slovakia's wine region, it was found that a third of vineyard area has been lost since the systemic change in the early 1990s. This is due to declining economic feasibility for what had been largely carried out by smallholders subsidised by the state but outside the cooperative farm system. As in much of the region, there were plenty of small private farms in operation prior to the restoration of capitalism. For abandoned vineyards in Slovakia, the outcome of systemic change is so far a conversion to conventional farming or urban area, leading to net biodiversity loss and more soil sealing (Lieskovský et al. 2013). Finally, a third major problem is that basically all studies converge on implying that reforestation (never even remotely close to reaching pre-industrial levels) requires social cataclysms to induce farmland abandonment, rather than democratisation.

Forests are nowhere near the main kinds of ecosystems in central and eastern Europe, and certainly not in other regions where socialist states once existed. In contrast to most of western Europe, for example, high-biodiversity grasslands make up a large portion of central and eastern European landscapes. Such grasslands have not exactly been prioritised for conservation after the systemic change. Not that they necessarily were under state socialism, except for the extensive system of nature preserves. In the USSR, 23 million hectares of steppe were destroyed to be converted into farmland (not unlike what happened in North America decades earlier), making what is now Kazakhstan into a centre of cereals production. It was the rapid economic downturn with the demise of the USSR that led to a partial reversal of farmland to grassland, only for the trend to be reversed again by the 2000s (Kraemer et al. 2015). In the Hungarian Danube-Tisza interfluve, after decades of grassland stability under the socialist state, about 15 per cent (400 km^2) was wiped out within 13 years (1987–99) through the plough, urban expansion and afforestation with mostly non-native trees. Further destruction ensued through 2008, though at a much lower pace of about 0.4 per cent (Biró et al. 2013). Similarly, in northern Croatia, between 1991 and 2012, most areas have witnessed a net loss of permanent grasslands despite the continuous decline of farming and pasture and widespread farm abandonment. A similar decline is reported in parts of Slovakia (Kanianska et al. 2014). More recently established grasslands, which account for most of the 21 per cent increase in grassland area since 1991, have developed at the expense of forests or farmland (Cvitanović et al. 2017). With such market-friendly management, the native grasslands of such regions may meet a fate worse than that of the North American tall-grass prairie. Unsurprisingly, conservationists are realising that there needs to be greater state intervention (sometimes without saying so directly), like providing grassland preservation incentives for owners or to introducing or enlarging protected areas.

The fate of other organisms is in part tied to what has happened to vegetation cover, but also to what people have resorted to do when in dire straits. Within the boundaries of the Czech Republic, Estonia, Mongolia, Romania and the Russian Federation, regime change brought about a drastic curtailment of funding for wildlife management and, with virtually unfettered poaching, a net temporary decline in large mammalian populations, with few exceptions. This happened even if major changes in land use eventuated in habitat improvement for some species. In the western part of Russia, abandonment of about 40 per cent of farmland has led to large tracts of early-phase forest and this may have helped some large mammalian species, like the wolf, brown bear and wild boar. But since the early 2000s several large mammal populations have rebounded (not so the Eurasian lynx, red deer and wild reindeer), so that the initial downturn at least from some species seems linked to the initial effects of the regime change (Bragina et al. 2015b), although the misery that change brought to most people has not substantively subsided. In a later study covering nine formerly state-socialist countries in central and eastern Europe, some of the same scientists (Bragina et al. 2018) show a net decline for the same large mammal species in Ukraine between the early 1980s and 2010s and mixed to positive trends for the rest. The picture is not entirely rosy for large mammals, nevertheless. There are doubtless some improvements for a few species, but these may also be due more to changing circumstances than constructive changes in environmental policy or more effective protection enforcement. To complicate matters, as with other regions of the world, it is unclear whether and to what degree large mammal migrations affect the reliability of counts that are based on national-level data.

Natural preserves and conservation efforts in general have gone mainly down the drain along with socialist states. With the fall of the USSR, the positive legacy of large tracts of protected lands (*zapovedniki*) has been undermined through the reduction of state support and increasing pressures on opening up *zapovedniki* to exploitation (Ostergren and Shvarts 2000). The break-up of the country into different countries has resulted in less, if not zero coordination of activities impacting shared ecosystems. The Aral Sea fisheries are one major example, where privatisation and division into separate national jurisdictions has accelerated an already declining fish population, including by overfishing and continued agrochemical pollution (Petr et al. 2004). Even environmental monitoring has become more challenging (Oldfield 2005, 94–5). Lioubimt-seva (2014), for example, reports that meteorological stations have rarefied or entirely ceased operations since the USSR was dissolved. This has also hindered weather data gathering on the Aral Sea. Water depth data are also harder to retrieve with the fall in the number of gauging stations (Laity 2008).

THE DISMAL ENVIRONMENTAL RECORD OF CAPITALISM

Available evidence does not support the view that state-socialist countries have had an environmental record that is worse than that of capitalist countries. This is so whichever kind of comparison one wants to choose. A cursory look into the worst impacts makes it obvious that businesses and capitalist governments can be shown to have caused many of the very worst accidents with the highest numbers of fatalities, especially when compared to Chernobyl.

With respect to the main destructive impacts at the planetary scale, state-socialist systems have never come close to the level of destructiveness of capitalist systems. Any worse record on atmospheric emissions was at most temporary and only for a couple of substances. The highest greenhouse gas emissions are squarely the business of capitalist countries. When one considers per capita CO_2 emissions the disparity is so large as to make even countries like the PRC look environmentally friendly. It needs to be kept in mind that often capitalist social system averages for some emissions are lower because most countries have relatively low GDP, smaller manufacturing and industrial farming sectors, and majorities suffering from inadequate resources to meet basic needs. Consequently such countries tend to emit less greenhouse gases and air pollutants, such as SO_2. State-socialist countries, on the other hand, included attempts to raise productivity levels to raise living standards as well as to develop the military in order to deter capitalist attacks and invasions. That such policies did not succeed in reaching the levels enjoyed in the wealthiest capitalist countries is the ruse by which anti-socialists distract attention away from the violent basis of global inequalities and from the exploitative sources of capital accumulation in liberal democracies (Amin 2018; Mies 1986).

The rapid industrialisation necessary to achieve the above-mentioned policy aims in state-socialist countries led to an initial dramatic spike, levelling and major to moderate declines in emissions. This should not be surprising, since levels of well-being correlate with the quantities of energy consumed, which is still derived overwhelmingly from burning fossil fuels (Schwartzman and Schwartzman 2019). Still, industrialisation led to unprecedented jumps in greenhouse gas emissions only in some of the state-socialist countries in the 1960s and 1970s, especially those that were initially largely based on peasant farming. In the meantime, emissions keep rising as production rises in already highly industrialised countries, whether state-socialist or capitalist. In some ways, at the planetary scale the influences of the USSR and then the Maoist PRC set limits on capitalist ruling classes' global appropriation of resources and environmental destruction in ways similarly noted regarding the mitigation of capitalist demands over workers in liberal democracies and the facilitation of national liberation movements' independence from colonial powers (Blackburn 1991, 192).

When the USSR and allied states transformed into liberal democracies some emissions fell, but within a decade they started climbing back up. In the PRC, on the other hand, a major change had happened by the late 1970s, leading eventually to the highest total emissions in the world by the 2000s. This is much more explanatory than any argument about capitalist superiority in energy efficiency or technological innovation. It would be a major feat if such techniques and technologies were environmentally low impacting (from extraction to consumption), as well as freely available and redistributed worldwide. What is omitted in the pro-capitalist argument of market-driven (or public–private generated) technological advances in lowering environmental impact is that attaining such technological heights, on capitalist terms, requires exploiting much of the planet for resources and thereby suppressing potentials for living standards improvements in most other countries, including capitalist ones.

If one is truly concerned about climate change, the cumulative and total greenhouse gases emitted into the atmosphere should be the principal focus of attention. Accordingly, the priority must be reforming capitalist systems away from fossil fuels and high-input industrial farming, starting with the highest-income capitalist countries, that is, beginning mainly with liberal democracies. This would have to include proscribing investments into other countries when they involve raising greenhouse gas emissions, including cattle ranching, mining and other sectors leading to deforestation and net increases in greenhouse gas emissions.

In other words, for a Global Green New Deal to be workable (Chomsky and Pollin 2020), the first step is political reforms in liberal democracies, which combine as the most responsible for total greenhouse gas emissions, currently and historically. After all, those countries are touted as bastions of representative government, free expression, individual rights, and the like. The fact that greenhouse gas emissions continue to rise unabated in spite of the popularity of green alternatives points to the failure of liberal democracies to address one of the biggest environmental problem facing humanity.

Pryde recognised long ago that environmental problems in the USSR, the usual target of animus, had by the 1960s 'become at least as bad as in the United States' (Pryde 1972, 171). This is in part false. The USSR was not even close to how terrible the US is environmentally, especially for the rest of the world. Nevertheless, even a supporter of liberal democratic capitalism felt compelled to concede parity even if not superiority, rather than inferiority, while insisting on faulty comparisons. So, the news is that state-socialist systems, given the evidence, are, to say the least, generally more environmentally constructive.

THE POVERTY OF COMPARISONS

Ultimately, as Peter J. Taylor (1987) pointed out some decades ago, explaining phenomena by means of comparisons is of doubtful legitimacy without examining how the interconnections and mutual transformations between the compared cases lead to those phenomena. The shaky nature of comparisons between countries becomes evident once it is realised that, especially since the intensification of global linkages in the 1970s, what happens in one country is traceable to if not contingent on what happens in another country (Dietz and Rosa 1994). This is without even considering the widely shifting conditions that characterise many countries, which may be spurred on by major catastrophes, economic downturns, internal rebellions, military invasions and other such.

Comparisons at the social system level of abstraction, as reproduced in part here, are a dubious undertaking. They entail effacing differences between countries within each social system. It is like pretending that all capitalist or socialist countries are or have been ruled through a single state or government. Hence, for example, per capita statistics should be reserved for comparisons between countries rather than social systems. This level of detailed comparison would also make more sense if the range of ecosystem variability within countries is also to be considered, but the task then becomes unnecessarily encyclopaedic.

There are additional, arguably even more serious shortcomings to the usual ways that comparisons are made. State-socialist countries are typically evaluated as if they had no inimical social structures to overcome or long-term environmental damage to face because of impacts from preceding capitalist systems. This omission of history becomes most disturbing in, among others, the cases of Angola, Cambodia, Laos, Mozambique, North Korea, Vietnam and especially Laos, the most bombed country in history (Boland 2017). The ecocides resulting from the wars inflicted on those countries are the direct responsibility of liberal democracies. Among the major horrific legacies are the destruction of forest ecosystems through extensive bombing raids, the mine fields laid out during the wars and the long-term effects on people of relentless attacks with chemical weapons (Schecter et al. 1995; Westing 1984; Zierler 2011). This is not even to touch upon the long-term ecological impacts of plantations, mines and other resource extraction processes imposed through colonialism in most of the countries that became, for a time, state-socialist. This applies to most of the above-mentioned countries along with others like Congo and Madagascar. These examples speak for themselves and flatly contradict the 'legacies of communism' storyline discussed more substantively in Chapter 5.

Just as problematic is the predominant selective comparisons of only some industrialised state-socialist countries with only the wealthiest and most militarily powerful liberal democracies, given the increasing and highly unfavourable financial and trade ties of the former to the latter. State-socialist countries were neither closed nor autarchical. People moved, commodities were exchanged and ideas were transmitted across boundaries, with restrictions imposed on such movements by both state-socialist and capitalist governments (Josephson et al. 2013, 107; Mignon Kirchhof and McNeill 2019, 8). In 1930s USSR, tens of thousands of tractors were imported from the US to help raise farming productivity. US companies like DuPont saw no difficulty in getting involved with and profiting from building the fertiliser industry and some of the major canal works to develop irrigation systems (Josephson et al. 2013, 96–7). Between 1929 and 1940 a quarter of industrial equipment and hundreds of thousands of machine tools were imported (Josephson et al. 2013, 84). Thus, capitalist industry also had at least a small hand in the environmental impacts of an industrialising USSR, unlike the much larger hand in the PRC since the late 1970s. But it also had a hand in progress over environmental monitoring and remediation. As in the case of the majority of countries, the equipment to accomplish those tasks had to be imported from the most powerful capitalist countries (Josephson et al. 2013, 200).

The wider context of socialist states has always been a capitalist world-system, whose existence is contingent on endless capital accumulation; that is, the primary motivation to produce anything is supposed to be to gain profits above all (Chase-Dunn 1982; Frey 2012). The capitalist world-system is predicated on forcing other societies into developing capitalist social structures (or risk perishing) and thereby enter into a subordinated relation to a capitalist core. The outcome is the establishment and reproduction of highly uneven development at all scales, meaning that most of humanity's life chances are contingent on the prerogatives of a handful of capitalists.

In conventional comparisons, countries that were formed or were transformed as an outcome of struggles against capitalism are being compared to the capitalist countries that invaded, colonised or indirectly ruled over those newly formed or transformed countries. Integration into the world capitalist economy is not optional and, arguably, neither is an industrialisation drive. Sustained ties with the capitalist world and raising productive capacity could be avoided only if one had been entirely uninterested in not just rebuilding, but also improving on the means to feed, clothe and house people after inheriting a low and extremely uneven level of production capacity, an economy structured around raw material and agricultural export (e.g. the USSR, China) or a debilitating plantation colonial economy (e.g. Benin, Cuba, Madagascar), and, for many countries, the direct devastation of two world wars with civil wars in between and/or afterwards (e.g. Bulgaria, China, Hungary, Laos, Mongolia, North Korea, Romania, USSR, Vietnam, Yugoslavia), causing millions of deaths and, economically, major labour shortages.

The last factor provided an additional incentive to mechanise production as quickly as possible instead of waiting for a demographic rebound. Industrialisation, on the other hand, was crucial to overcoming chronic food shortages in countries where socialist states arose. The food production challenges had little to do with state socialism. In the case of the USSR, for example, most of the productive farmland had been wrecked by the invading Nazis, resulting in a 40 per cent contraction of agricultural production, followed by a protracted and extensive drought in some of the otherwise most agriculturally productive regions (Josephson et al. 2013, 113). It should hardly be surprising that USSR's postwar economic trajectory resembled that of Japan, with similar percentages of international trade patterns relative to national economic output (Sanchez-Sibony 2014, 4–5).

All state-socialist countries faced externally imposed challenges, while tackling internal strife and attempting to bring about major social change towards equality in often oppressive patriarchal peasant or pastoralist systems. The horrific tragedies and, in some cases (especially the USSR under Stalin), mass killings and grinding political repression are unjustifiable, but the larger context must also be understood to see what needs to be done to avoid those horrors in future. Some state-socialist systems emerged through the establishment of political independence after protracted war but were then trapped in neocolonial dependence and interventionist pressures from former colonial powers, such that building self-sufficiency was stymied, and even more any chances for industrialisation (e.g. Burkina Faso, Congo, Ethiopia, Madagascar). Some extricated themselves from a neocolonial or colonial dictatorship and started substantive independence under terrible social conditions (e.g. Cuba, Mongolia), if not absolutely devastated environmental circumstances (e.g. Kampuchea, Laos, North Korea, Vietnam). There were thus intense pressures to parry from capitalist countries. Rapid industrialisation or the promotion of self-sufficiency was eminently reasonable, given the experience of military invasions from liberal democracies and other capitalist powers, along with military encirclement and constant threats after repelling those invasions. In some cases, foreign powers and internal enemies waged all-out war from independence onwards, such that any effective implementation of socialist policies was altogether precluded (e.g. Angola, Mozambique). Briefly put, socialist states existed under a state of siege from their inception.

When siege intensity lessened, the pressures were economic and not terribly different from the sort experienced as neocolonialism in most former colonies at formal independence. There were and still are embargoes and other ways through which state-socialist economies have been undermined by concerted efforts through various international institutions like the IMF and the Trilateral Commission. As soon as the opportunity arose, capitalists (especially those from liberal democracies) took advantage by setting up shop or establishing joint ventures within state-socialist countries to exploit highly skilled labour for a pittance (Frank 1977; Pryde 1972, 159). Environmental destruction in state-socialist countries resulted from these capitalist pressures and interlinkages, too, not just from some allegedly misguided domestic policies or socialism per se.

In the face of such enormous disadvantages, state socialism was in many instances an overall success. It lay the groundwork for and to a large degree achieved reductions in social inequalities and improvements in living standards and environmental practices, while fighting off colonialism or imperialism from core capitalist countries and trying to overcome internal reactionary resistance. Those accomplishments should not be tossed aside and consigned to oblivion. One does not jettison voting rights, progress against slavery systems or advances made in environmental justice in liberal democracies on account of the multiple genocides and racism that subtends the very core of those political systems.

In other words, for the story to hold that state socialism is worse on the environment, historical and international contexts must be entirely or partially eviscerated. The conventional narrative is an attempt to invert reality, where official pronouncements or stated policies are taken at face value only when they show 'the intention of challenging and eventually superceding the West' (Oldfield 2005, 33; Chu 2018). Institutional expressions regarding social equality, building socialism, raising living standards and improving the relationship to the environment are instead met with scepticism (cf. Pryde 1972, 136). In this manner, as propaganda is conflated for reality when convenient, socialist states are made into caricatures: easy targets of scorn or dismissal. It is the triumph of politically innocuous critique and of socially irrelevant irreverence to 'really existing socialism', all to the current ruling classes' merriment.

When the countries being compared have developed intense connections over time, the comparison can even be disinforming on environmental change. During the socialist state period in Hungary, for example, widespread soil acidification problems are traceable to heavy use of agrochemicals, like ammoniacal nitrate, but the problem could have been averted or at least reduced by refraining from producing the likes of sausages, fruits and vegetables for Western markets, especially West Germany and Austria, to fulfil loan repayment schedules imposed by Western financial institutions (Engel-Di Mauro 2002). In effect, the more an economy is integrated with international capital flows, the higher the ecological footprint (Figge et al. 2017). Accordingly, China, as a country embodying the main contradictions of the world capitalist economy (Frey 2012; Li 2016), should be expected to have a worsening environmental record. As Peters et al. (2012) have shown, the wealthiest countries (largely liberal democracies) can overconsume fossil fuels and spew out the most greenhouse gases by appropriating natural wealth (e.g. fossil fuels) from the rest of the world, where there is chronic underconsumption relative to fossil fuel production.

There are additional problems to consider related to diverse and changing biophysical contexts. For example, a short-grass prairie does not respond in the same way to the same kind of farming impacts as a mixed deciduous forest. Once trees are felled they may regrow as secondary forest, not grasses. Likewise, grasses that are ploughed may also be restored and forests would be unlikely to appear even if planted. Soil types are not going to suddenly change because a mixed deciduous forest is gone. Such soils will tend to retain the main characteristics (often on the more acidic side, for example) even when forests do not return as a result of constant ploughing. Some soils are more erodible or have more material to erode than other soils even within the same kind of ecosystem. Moreover, comparisons of forest ecosystems even within the same region can be suspect. In a study conducted recently in the Urals it was found that plant species in taiga biomes are more sensitive to crude oil contamination than those making up mixed conifer-deciduous forests (Buzmakov et al. 2019). The same intensity of crude oil contamination, in other words, will be deadlier for some forests than others. These kinds of differences are often papered over, which can lead to exaggerating or underestimating damage.

To obviate misleading comparisons of countries' environmental records, both biophysical and social contexts need to be examined along with interlinkages across places. To some extent this is accomplished in Chapter 4 by looking more closely into the ecosocial contexts and histories of the USSR, China and Cuba.

4

Environmental Impacts in Context

As discussed in Chapter 1, state socialism is more akin to a transitory system that can go the way of socialism or of some guise of capitalism, if not of an entirely different type of oppressive system aside from capitalism. State socialism is a contradictory set of conditions where struggles continue for socialism, characterised by internal strife among contending socialist and anti-socialist forces, under constant threats of annihilation by core capitalist states in a wider global context dominated by capitalist relations. This transitory and internally contradictory process situated in mortally antagonistic relations of power extends to and shapes the relationship of socialist states with the environment.

On the one hand, socialist revolutionaries attempt to develop the material conditions of socialism by raising everyone's well-being, while parrying deadly attacks from within and from abroad. This in itself already poses a contradiction, since warfare denies the possibility of considering everyone's well-being. Internal conflict includes conflicts within and among socialist movements relative to the best course of action to achieve similar or the same short- and long-term objectives. On the other hand, as part of improving people's lives and establishing the material basis for a socialist future, ecological conditions must be protected with the same vigour. At stake are public health (or workers' life prospects) and the ecological and environmental foundations for the resources needed to build socialism.

The contradictions and social struggles are therefore multiple, and simultaneously social and ecological. Navigating them was and is a challenge to socialism generally and socialist states are unexceptional in this. The issue here is how under such contradictions and struggles socialist states fared relative to the biophysical environment and what they can offer for future socialist prospects. It will be shown that, as socialist states, they fared much better than we are usually told and that they provide examples on which ecosocialism can be built. However, when looking into the fate of the relationship between state socialism and the environment the starting position should never be downplayed. The historical conjuncture (prevailing global relations of power) and the social contexts of successful socialist revolutions deeply inflect potentials for socialism and relations with the rest of nature.

All societies where revolutions succeeded were characterised by privation and tenuous living conditions for the majority. Most of those societies were ravaged by centuries of colonial depredation or by recent or sustained devastating wars. Invariably there were destructive capitalist legacies and tendencies, both ecological and social, that had to be redressed and shaped the possibilities of what could be achieved. Under such harsh circumstances, environmental concerns must be weighed against the necessity of rapidly building, often for the first time or almost from scratch, the foundations to enable coverage of basic survival needs while also improving people's living standards. These include establishing reliable food, water and medicinal supplies, adequate housing and sanitary infrastructure, effective health provision and transport, among other ways of meeting basic human needs and of improving living conditions that by and large societies undergoing socialist revolutions lacked historically. This balancing act between ecological relations and improvements in living conditions occurred under the constant military and economic pressures of liberal democracies, directly or indirectly. At the same time it must be underlined that it was not by conquest and colonialism that socialist states achieved or tried to achieve such balance. The destructive environmental and social outcomes from developing industries to improve material conditions were consequently meted out overwhelmingly within state-socialist countries' borders (Weiner 2017).

To elucidate on the contexts that shaped the environmental and ecological outcomes of state socialism I rely on three case studies. They are selected on the basis of their extent of global influence, substantive social and ecological/environmental differences and the author's relative familiarity and background with those countries. The first is the USSR, in accordance with historical sequence, followed by China and Cuba. The USSR set the pace for state socialism worldwide and, so far, exemplifies the longest period of state-socialist environmental impact. The net environmental effects were more positive than negative, even when considering the calamities for which the USSR has been made infamous in the press and within the academy. In contrast, the PRC is a showcase for the effects of an arrested state-socialist path redirected to a form of one-party socialist rule over a capitalist economy. The environmental repercussions have been dire, but not as hopeless as often depicted. It may resemble what has happened in countries like Laos and Vietnam. Cuba, on the other hand, presents a unique case of state-socialist survival, largely independent of both the USSR and China, and an ecological model for the rest of the world. In some ways, Cuba exemplifies how a socialist state can lay the groundwork for an ecologically sustainable socialism or ecosocialism.

A further elaboration on the institutional set-up that affected state-socialist environmental practices can help guide the interpretation of the ecological effects of state socialism, picking up from the discussion in Chapter 1 regarding what characterises state socialism (see also O'Connor 1998, 255–65). Because politics and economy were overtly linked, instead of being masked as in capitalist societies, environmental issues were immediately a matter of direct political struggle over how resources are to be allocated, used and even defined (e.g. what are to be considered legitimate sources of energy and using what criteria). Most enterprises answered to one or another form of centralised administrative decision-making body and ultimately to a communist party's central authority. These characteristics also make the party directly responsible for social and environmental outcomes, so that an entire political system's legitimacy rides on policy results.

For example, centralised price determination could be directed at mitigating the consumption of raw materials and wastefulness in general. Potentially, it could also redirect production towards ecologically more constructive ends (Goldman 1972, 273–83; Josephson et al. 2013, 214; Oldfield 2005, 25; Pryde 1972, 153). That it was not always so and, given similar or comparatively poorer records in the most powerful capitalist countries (think of the Persian Gulf monarchies or the US, especially in terms of global impacts), should be grounds for further investigation into negatively influential factors and ways to combat them in future if one is interested in struggling for socialism rather than totalising dismissals of socialist states.

A fundamental flaw, though, was or is the lack of direct account-ability of different institutions, from enterprise to the central state decision-making organs, to the people supposedly being represented, allegedly the working class. Worker control (e.g. through council/ soviet communism, workplace democracy) can be of decisive help in identifying environmental harm and prioritising its prevention, especially when human health is in question. Still, worker control is no guarantee of an environmentally friendly outcome, as fighting for better working conditions or shorter working days has no particu-larly ecological content. There is therefore another aspect that needs consideration, which is the level of ecological sensibility in society as well as the incentives and consequences of ecologically sustaina-ble practices and environmentalism. Perhaps to state the obvious, the sensibility and incentives emerge, in large measure, from the lived material conditions of deteriorating ecological relations or physical environments. In state-socialist systems the level of ecological sensi-bility and incentives for ecologically constructive actions were mainly shaped by the results of struggles within party structures, constituted in part by environmentalists (see the discussion on green Bolshevism below) and by pressures on the party apparatus from wider sections of society, including from scientific communities and volunteer associa-tions, and inimical forces within and outside a state-socialist country.

The USSR's decision-making structures set the template for other state-socialist systems, so it is useful to consider it in elucidating envi-ronmental policies and practices for state socialism broadly. In the USSR, republic and overarching All-Union Ministries governed tens of thousands of firms. Short-to long-term planning for each ministry occurred through the Gosplan (State Planning Committee), whose national plans were subjected to the scrutiny of the Supreme Soviet, the legislative branch representing the republics and deciding over all planning projects. Planning included resource use and conservation measures, which each ministry drew up relative to their area of com-petence. Conformity to environmental regulations was enforced from within each ministry, but also from the millions of inspectors in the People's Control Commission, which was established in 1920 (as the Workers' and Peasants' Inspectorate) to serve as supervisory control over administrations and enterprises. The ministries in turn were under the responsibility of the Council of Ministers, whose members were elected by the Supreme Soviet. Eventually the executive organs of the Communist Party of the USSR, elected through party congresses, had the final say (depending on the period, this was the Central Committee or its more restricted Politburo component). The USSR's decision-making structure was not transplanted whole in all state-socialist countries, but many of its facets were introduced and modified to suit national contexts. Importantly, state office workers (mainly bureaucrats) did not necessarily have to be Communist Party members. State apparatuses and bureaucracies were never solely in the hands of a single party.

This decision-making structure facilitates coordination of activities across different levels, from national to production unit level (e.g. mine, farm, office, factory). Scientific research and higher education institutions provided expertise-based information and advice for different specialised departments and ministries, which conveyed the findings and recommendations to central planners to draw up national plans. Environmental monitoring and information flow were standardised as part of scientific work and the data were fed directly into the ministries responsible for different branches of the economy. In addition, in wider society, volunteer associations (e.g. naturalists) and scientific establishments concerned with biological and health issues conducted independent studies, monitoring and campaigns to educate or sensitise the public to environmental issues. Destructive environmental impacts could be easily tracked at the highest decision-making levels through a variety of channels (Ziegler 1980).

The main obstacles to avoiding environmentally negative impacts or to prioritising ecological issues, aside from there being no ultimate accountability to workers at the workplace level, were the investment priorities and rewarding system, deformed as well by external international pressures. Central state authorities, within and beyond the party, privileged military defence and heavy industry, distributed funds to enterprises mainly according to productivity (production targets, fulfilling plans) and rewarded enterprise management with bonuses and privileges. This kind of framework favours managers' enterprise-oriented or sectoral self-promotion at the expense of wider social interests and ecological concerns. The problem would persist even if economic activities were organised in worker-run cooperatives. Anything impeding the fulfilment of production plans could compromise the status and viability of a production unit.

It is important to keep in mind that interministerial conflict was rife not only over resource allocation, but also over environmental protection, as well as regulations and their enforcement. Hence there were and are multiple sources and forms of environmental information reaching the central decision-making bodies (Ziegler 1980). As a result environmental records could never be entirely distorted or suppressed, not even to the wider public. At the very least, central decision-making offices would have a clear picture of environmental problems, even if sometimes with delay when lower-ranking administrators slowed down the data-collection process or falsified data for personal gain. Nevertheless, coordinating such wide-ranging interests relative to environmental policies in a state-socialist system rests mightily on central party authorities, at once a strength and weakness technically: a strength in mobilising people and resources to address environmental problems; a weakness in being potentially overwhelmed by too much information and too many sets of relations to handle. Ultimately, the overall directions of coordination efforts result mainly from struggles within those centralised political structures, even as they are heavily affected by forces external to them.

Exacerbating the problem is that work and institutions dedicated to environmental protection and ecologically constructive economic activity received considerably fewer material incentives. Central party authorities, at any administrative level, also worsen prospects for ecological sustainability when repressing pro-socialist environmentalist forces and by suppressing environmental information when it may, over the short term, undermine socialist state legitimacy. But, just as in the rest of society, these authorities were (and are) also divided relative to priorities, objectives, strategies and other such considerations. As in any social system, state socialism is imbued with relations of power and social struggles, except that their outcomes can, as already remarked, lead to building or strengthening actual socialism or degenerate into some form of capitalism or other kind of authoritarian system (North Korean dynastic rule could be viewed as an example). The outcome is a varied history of environmental impacts, with some horrific failures interspersed among spectacular as well as many muted successes.

THE USSR: CREATING MASS ECOLOGICAL CONSCIOUSNESS

By far the greatest critical scrutiny has been showered on the USSR when it comes to socialism and the environment. The first of what turned out to be a sensationalist account appeared to a Western public in the early 1980s in the form of a translated samizdat (self-published manuscript). It was authored by a Boris Komarov, Ze'ev Wolfson's pseudonym. The samizdat was smuggled to the West and quickly seized upon in liberal democracies as a chance to whip up more anti-communism, this time on environmentalist grounds. The work was deeply flawed, argumentatively and empirically, as Marshall Goldman remarked in the foreword to that book. Wolfson had skipped the critical assessments levelled from within government institutions as well as policy successes in reversing or preventing ecological damage. In spite of Marshall's warnings, major scholars have reproduced the flawed arguments and problematic data, which were used to charge the USSR with 'ecocide' (e.g. Feshbach and Friendly 1993; Pryde 1986). The high-ranking geographer Wolfson would later migrate to Israel, where he strangely never took issue with the environmental devastation there (Tal 2002), much less with an annihilationist settler colonial system (Finkelstein 2003). He instead turned his attention again to the then former USSR, finding, to his chagrin, that the free market is even worse (Wolfson 1994). But by that time any critique of capitalism had great difficulty gaining any traction. Wolfson's take may have contrasted official USSR government statements on environmental degradation, but it also amplified the already loud and then increasingly louder fantasies equating freedom, markets and democracy with a healthy environment.

What materialised after Wolfson's samizdat was much more influential than any smuggled unauthorised work. What Komarov decried in the 1970s paled in comparison to the Chernobyl accident and the shrinking Aral Sea, the most popularised, prominent examples of environmental damage. Enduring ecological harms they certainly are, but they have been easily subordinated to what could be considered an environmentalist equivalent of a red scare, where even reforming capitalism through a 'Green New Deal' is too socialist. It is for this reason that I discuss those three cases of environmental destruction in somewhat greater detail. They demonstrate how much historical and international capitalist context matters and how the very worst imputed to the USSR conceals a more widespread opposite tendency of sensible environmental policies and practices, certainly compared to those suffusing the new 'post-socialist' systems. Buried in the very texts where the USSR is indicted as a failure in environmental protection are examples of cutting-edge biological conservation practices through ecological preserves (*zapovedniki*), successes in pollution abatement, lasting improvements through soil conservation, afforestation and reforestation, mass diffusion of environmental education and ecological sensitisation and some of the most advanced environmental research and monitoring programmes worldwide. In terms of positive global contributions, it is regrettably much forgotten how repeated efforts by the USSR to link disarmament to international environmental treaties were rebuffed by the US and allied states (Josephson et al. 2013, 196–7).

Czarist Legacies

First, however, a few words on Czarist legacies are in order. The USSR did not emerge out of a cocoon. Especially in western Russia, forests and soils came under severe pressure with industrialisation, especially in the rising centres of factory production in major cities and of fossil fuel resource extraction, like Baku for oil and the Donbass for coal. Commercially motivated hunting and logging led to rapid falls in the populations of bears, beavers, elk, fox, hare, mink, sable, seals, wild boar and wolves, among other species that were increasingly threatened with at least local extinction if not locally or regionally extinguished (Josephson et al. 2013, 41; Pryde 1972, 9–11).

As one of the main culprits of species loss, deforestation was rife in part because of its economic linking to the vast and rapid forest losses in North America, among the many destructive outcomes of US and Canadian settler colonial expansion (Wynn 2007). By the late 1800s, US and Canadian business demand for timber was mainly fulfilled by Czarist Russia, which had turned into a main timber exporter (Josephson et al. 2013, 121). A similar process of Russia being converted into a raw material supplier for North American settler colonial empires occurred with the fur trade, which devastated beaver and other fur-bearing animals as well as Indigenous communities in both Russia and North America.

Forest losses were also due to Czarist economic policies and were particularly intense on private estates (Josephson et al. 2013, 31). For instance, the now much deforested and polluted Kola Peninsula in the Arctic was already an object of railway infrastructure development and resource exploitation before 1917 and using forced labour. Under Stalin, by the 1930s Czarist policy and brutality was much expanded, not newly introduced, and it even involved some of the same people at the helm (Bruno 2016, 12–13).

In the newly conquered lands of Central Asia, starting in the 1860s, the Czarist regime displaced millions of nomadic herder peoples and, by severely undermining transhumance, increased susceptibility to famine. Ploughing the steppe contributed to increasing the chances of pasture and cropland failures during droughts. In 1855–6, Kazakh herders lost as much as 70 per cent of cattle due to such calamity. The worst came during 1930–3, due to grossly negligent USSR policies, but it is also true that Czarist government policy led to half a million deaths in the 1891 famine, even though enough grain was available. Famines and the economic maldistribution that exacerbated them were a periodic feature of Russian life under the Czar that would only be surmounted in the 1950s, thanks to the advances made possible by succeeding USSR governments.

Much of the settler colonial effort amounted to raising existing cotton production to make the region a cotton basket, introducing large monocultures, massively expanding irrigation and building railways mainly linking resource extraction and commercial centres. By World War I the Russian Empire had become one of the largest exporters of cotton in the world, much to the detriment of local ecosystems. Waterlogging and soil salinisation became a widespread problem, as did the malaria outbreaks related to the lager swampland area created (Cameron 2018; White 2013).

With the 1861 serfdom reforms, peasants were granted titles to small land parcels. These were mainly in sloping zones with relatively thin soils poor in nutrients. This policy unleashed the largest historical expansion of cultivated area, covering much of the southern and south-eastern Russian Plain. The result, combined with inadequate means and techniques at peasants' disposal, was severe gully erosion, the sort that mars a landscape for the long term and is difficult to remediate. Effects were felt downslope in the flow impairment if not clogging up of small rivers and in the build-up of sediment along the valley bottoms (Golosov and Belyaev 2013; Sidorchuk and Golosov 2003).

There were attempts in some quarters, by the late nineteenth century, to lessen if not stop the ecological destructiveness of the Czarist Empire. These culminated in the introduction of the first *zapovedniki* in the 1890s, or nature reserves equipped with scientific research stations, and an initiative to establish experimental stations to help devise soil conservation measures more effectively. Much was contingent on the intensity of commercial pressures. More importantly, nature protection, such as there was to limited degrees under the Czar, was conceptualised in terms of long-term reproducible yield of whatever resources, mainly timber, suited capitalist prerogatives. It was not an ecosystem that was to be preserved, but bits and pieces of it: those that made money or served the state's needs of the day (not that the two are incompatible, but they might not always coincide over the short term).

To some extent, especially among biologists and ecologists, a strong environmental sensibility had evolved since at least the 1860s and that had been frustrated by the poor environmental record of the Czarist regime. Aside from forest losses and soil destruction, biodiversity was threatened as many species were hunted to extinction or disappeared through habitat loss. Many more species were threatened by exploitative ravages, such as the above-mentioned fur trade. The destructive environmental impacts were associated with industrialisation, economic subordination to core capitalist countries and the historical imperialistic push eastwards into Siberia, Central Asia and elsewhere.

Early Bolshevik Environmentalism

As soon as the USSR was established, even in the midst of a most bloody civil war, the Bolshevik government introduced much environmental legislation and restrictions on industrial output were introduced. This was decades before they were introduced in countries of much greater economic means (Goldman 1972). Examples of early Bolshevik environmentalism and marked improvements in environmental protection measures abound. Some policies were also innovative and unprecedented. Again, during civil war, in 1921 the Bolshevik government put together the first agro-meteorological service in the world (Elie 2018, 83). This would later also be useful in climatological monitoring. A 1919 decree on water conservation aimed at preventing contamination of reservoirs from sewage established a water protection agency (Goldman 1972, 17). Wastewater regulation came into effect in 1923 (Pryde 1972, 137).[1] Green spaces and mass transport were already planned or in the works during the 1920s for urban centres. This began a decades-long effort to improve the quality of life in cities and it is why there is usually much more ecologically sustainable infrastructure now in the former USSR (Josephson et al. 2013, 91–3). Urban-planning efforts culminated in the materialisation of ample city parks, green belts, no suburban sprawl and general cleanliness in public areas (Pryde 1972, 156).

Conservation and forest and wildlife preservation were already policy priorities by 1917 (Goldman 1972, 16–18). Shortly after the civil war, between 1925 and 1929, the *zapovedniki* area was quadrupled from one million to four million hectares. This is one reason that most conservationists and naturalists initially had a favourable

---

1   Comparable laws did not see the light of day in the US until 1948 with the Federal Water Pollution Control Act (see 'History of the Clean Water Act', www.epa.gov/laws-regulations/history-clean-water-act, accessed 24 December 2020).

view of the October Revolution and many continued to do so even during more arduous times under the Stalin government (Josephson et al. 2013, 66; Weiner 1999). They saw it as the chance for new and existing policies, for which they had long fought, to be finally enforced effectively. They also saw in the revolution the potential for expanding and intensifying conservation programmes. But the main exponents of conservation sought the support of the Provisional Government, rather than the Soviets. They were aghast at the destruction of *zapovedniki* resulting from war, as well as, sometimes, from peasant takeovers of large estates and private parks (Suing and Dedaj 2018; Weiner 1988, 18–20).

A main problem, aside from a lack of involvement or membership in leftist parties, was that mainstream conservationists did not take their concerns to the people at large except in efforts for a general defence of science from emerging challenges by Proletkul't and other movements aiming to replace existing science with a proletarian science (Gare 1993; Sheehan 1985). The Proletkul't movement was formed in 1918 by Bolsheviks who believed that society in its entirety had to be transformed, including culture, and that promoting worker control of industry had to take precedence as well as the development of constructive relations with the rest of nature. Lenin was among the biting critics of Proletkul't, but he nevertheless did not try to stifle them. It was thanks to the scientific and Marxist outlook of influential segments of the Bolshevik Party, particularly leaders like Lenin and Lunacharsky, that conservationists found their views vigorously supported by the new government. Forests, surface waters and underground mineral resources were all nationalised and brought under conservation policies. Even under war conditions and threats to the very survival of the new state, efforts continued with protecting forests and new nature protection decrees were introduced (Pryde 1972, 14–15).

This was accomplished by actively seeking the collaboration even of eminent scientists who bore little to no sympathies for the revolution or communism, like V.I. Vernadskii. Under early Bolshevik government, environmentalism had a sort of roaring twenties, the likes of which no state-based society had ever witnessed. The form of environmentalism espoused by the likes of Lenin may not have diverged appreciably from the conservationism in the industrialised capitalist countries, but it was not the sort that led to mass displacements or appropriation of land to protect 'national interests'. And in that kind of conservationism more radical approaches could thrive, such as those of the Proletkul't movement. It was also under such favourable conditions for environmentalism that ecology thrived. Ecologists in the USSR were at the cutting edge of research and theoretical advances, including the approaches of phytosociology, community ecology and food webs and energy transfers, the last of these developed by Stanchinskii (Gare 1993; Suing and Dedaj 2018; Weiner 1988, 285, fn. 15).

Ecologists were gaining major influence over economic policies as well. In 1926 the Interagency State Committee for the Protection of Nature was instituted to vet all matters related to resource management and they even had veto power for plans deigned too environmentally destructive. This inchoate form of environmental impact assessment had no precedent and did not exist in any other state in the world. None of this was enough to save ecologists from marginalisation by the early 1930s, when many of them started to clash with the Bolshevik mainstream over the Five-Year Plan, owing to aspects that lifted protections for some *zapovedniki* and undermined other ecosystems.

Mass Environmental Sensitisation in the USSR

The heyday of green Bolshevism may have been short-lived, but it was never eradicated. Approaches developed and diffused during that period were elaborated upon and expanded in different ways and under diverse guises until the end of the USSR (Yanitsky 2012; Weiner 1999). Even under the repressive and murderous Stalin government there existed independent, critical movements for environmental protection, mainly led by scientists. To a large extent this resulted from the Bolsheviks' privileging of science as integral to state planning, policies and education. Though government censorship did impair scientific research intermittently over the 1930s and 1950s, mainly in genetics, scientific research was promoted and well funded and scientists had prominent political and wider social

influence (Weiner 2006). The latter was important in spreading ecological sensibility in the public and in struggles for environmental
protection, often against economic planners and technocrats.

A main conduit of scientific outreach and mass public involvement in environmental matters and struggles was the All-Russian
Society for the Protection of Nature (VOOP). It was founded in 1924
by prominent field biologists and enjoyed much popularity, boasting
tens of thousands of members less than a decade later, declining to a
few thousand under the repressive rule of the Stalin faction (Pryde
1972, 20–1; Weiner 1999, 37–8). By the 1970s, however, the society
boasted more than 25 million members and then 37 million by 1987,
about 13 per cent of the USSR's total population of the day (Foster
2015; Kelley et al. 1976, 178). The reach of environmental education
and ecological sensitisation was truly vast. In their openly critical and
politically independent engagement with the government, VOOP was
complemented by the Moscow Society of Naturalists, the Moscow
section of the Geographical Society of the USSR, the All-Union
Botanical Society, and several volunteer naturalist associations. It
was through the many projects and public outreach programmes
of these organisations that tens of thousands of people gained environmental education through volunteer work. At some universities
students formed nature protection groups like the Nature Protection
Brigade at Moscow State University, founded in 1960. Environmental
concerns were not circumscribed to scientific communities and the
many thousands of naturalists in the wider public. There was direct
political support from authorities at the republic and oblast (pro-
vincial) administrative levels as well as from portions of the official
press. This attests to a widespread sensitivity to environmental issues
throughout society, including governing institutions.

These organisations would write complaint letters directly to the
Politburo, publish letters of protest through the mainstream press,
carry out street protests to pressure authorities and even set up mock
show trials of high-ranking officials on a range of environmental questions, all without repressive consequences. This is how, for
example, the secret police were successfully taken to task in 1948
for logging in a protected area and how the rescinding of the inviolability status of several *zapovedniki* under Stalin was reversed and

full protection restored in 1954 (Weiner 2006). Mass mobilisation and campaigning were also behind stymieing technocrats' plans, on multiple occasions, to locate pulp mills near Lake Baikal (Kelley 1976) and to shift river flow in Siberia to undo damage to the Aral Sea. Several other large-scale projects were similarly blocked, as in the case of a hydroelectricity project that would have sacrificed Lake Sevan in Armenia (Breyfogle 2015; Pryde 1972, 122; Weiner 1999).

Ecological sensibility reached the highest echelons of power. Enterprise managers and associated economic ministries were repeatedly and scathingly criticised from within the upper chamber of the Supreme Soviet over inadequate pollution control measures. Throughout the history of the USSR, pressure to comply with environmental norms was also applied to economic planners from the Ministry of Public Health and from the People's Control Commission, with its millions of independent inspectors (Ziegler 1980). By the 1980s hundreds of environmentalist groups formed and successfully defeated major new environmentally destructive large-scale projects, and imposed the closure or retrofitting of polluting industrial plants (DeBardeleben 1992). The tide was finally turning against the technocratic economic planners and engineers and in favour of developing an ecologically based socialism, until the demise of the USSR halted the process in its tracks and partially rechanneled environmentalist movements towards nationalism, as attested in environmentalism's current parlous state of affairs in the countries formerly within the USSR (Elie and Coumel 2013; Newell and Henry 2017; Yanitsky 2012).

The USSR's Ecological Preserves (Zapovedniki)

It was especially the preservation of nature that animated much of the environmentalist opposition to government economic planners within different levels of government as well as among scientific and naturalist volunteer associations. The *zapovednik* or ecological preserve was a central pillar of environmentalism within and beyond the state. As described above, the first protected areas or *zapovedniki* designated for scientific study were instituted by scientific societies by the 1890s, but they lacked substantive legal protection and were designed to support commercial maximum-yield prerogatives. *Zapovedniki* 'were imagined as baselines (etalony) of healthy natural communities against which changes in surrounding, once-similar but human-affected areas could be compared' (Weiner 2006, 531). But only under the Bolshevik government did the *zapovednik* gain policy priority, firm legal backing and permanent inviolability status. *Zapovedniki* are a hallmark of the USSR's positive ecological impacts. It was soon after the revolution, in 1917, that the Bolshevik government established the first national *zapovednik*. It was also the first of its kind in setting aside large, protected areas for scientific purposes, as ecological exemplars. In a 1921 decree (reaffirmed in 1960), *zapovedniki* were to be reserved for ecosystem study and withdrawn permanently from economic use. By the 1940s the number of *zapovedniki* was expanded to 40, encompassing seven million hectares. Another expansion was instituted in the 1950s, with 12.5 million hectares spread over 128 *zapovedniki*. After a contraction in the 1960s, more such protected areas were decreed until they amounted to 150 areas, or 20 million hectares. Including surface waters, *zapovedniki* total area reached slightly more than 30 million hectares. Sizes varied considerably, from a couple of hundred to several million hectares. Half of the *zapovedniki* were located west of the Urals and in the Caucasus and they tended to be smaller than in less populated regions like Siberia (Colwell et al. 1997; Ostergren and Jacques 2002, 111–12).

As ecosystem models they were used as controls relative to heavily impacted areas to help pinpoint environmental problems more precisely and to develop ideas for more sustainable economic practices. Equipped with monitoring stations, these areas were subjected yearly to season-specific studies, generating an unmatched wealth of long-term ecological data. This includes the possibility of modelling different kinds of human impacts, the effects of climate change or fire disturbance on biological communities and the relationship between reserve size and geometry on the survival of different species. Eventually the geographical distribution of *zapovedniki* represented the major bioregions within the USSR's boundaries (Colwell et al. 1997).

An important aspect of early USSR policy was to entrust conservation mainly to the People's Commissariat for Education (Narkompros) so that natural resources would not be under the direct control of state organs with economic interests in exploitation. However, it was the People's Commissariat of Agriculture (Narkomzem) that was able to arrogate to itself responsibilities for hunting regulations and enforcement and the creation and maintenance of related *zapovedniki*, sometimes undermining conservation efforts (Weiner 1988, 24–30). During the hardships after the war and with the advent of the New Economic Policy, *zapovedniki* had to be subsidised by means of some resource extraction. This set a precedent that would later be usurped by some factions, especially in times of declining general prosperity (Weiner 1988, 39).

Conservation continued even at the height of the 'Great Plan for the Transformation of Nature' (1948–53), typically indicated as paragon of Stalinist excesses. Often the important background to this is omitted, which was in the main a response to the famine of 1946 and part of efforts to recover industrial capacity from the ravages of the world capitalist powers' conflagration, including especially Axis-Alliance invasions (Josephson et al. 2013, 84). The plan entailed an attempt to reduce the rough climate conditions (prone to extended droughts) in the southern and Central Asian portions of the USSR. This was tried by expanding irrigation canals, building artificial lakes and planting trees to create shelter belts. Throughout this period the number of *zapovedniki* was increased from 37 in 1937 to 128 in 1951, with the total area expanded from 7.14 to 12.5 million hectares (Pryde 1972, 51). What is more, millions of hectares of forest were set aside because of their importance to the viability of hydroelectric dams. Clear-cutting could have resulted in much soil erosion, eventually silting up rivers and impoundments and reducing the water flow necessary to generate electricity. To achieve this the Politburo repeatedly fought off the Ministry of Heavy Industry and Stalin mainly sided with conservationists on this issue (Brain 2011). The result was that, 'from 1948 to 1953, the number of trees planted exceeded that of those planted during the previous 250 years of forestry history' (Josephson et al. 2013, 120).

All this may not have been entirely successful and, in part, damage ensued. Some species like the Turan tiger in the Aral Sea area disappeared (White 2013). At the same time, thanks to measures taken by succeeding USSR governments, most species, such as the saiga and the sturgeon, were saved from extinction and, since capitalism took over, are now threatened with or nearing extinction (Bekenov et al. 1998; Milner-Gulland et al. 2003; Secor et al. 2000). In some cases the motivations for saving a species coincided with the interests of sections of some economic ministries. Sturgeon life, for instance, coincided with the economic importance of caviar. The issue of saving the sturgeon brought together local citizenry, the caviar industry and the State Fisheries Committee to mount a successful offensive against the building of a hydroelectric dam on the Volga river (Josephson et al. 2013, 171). But most of the efforts at biodiversity conservation were not reducible to particular economic interests. They were part of genuine environmental concerns within and outside governing institutions (Weiner 1999).

Pressures increased from economic planners to reduce the number and total area of the *zapovedniki* to make room for industrial uses. It was after the Stalin administration that the peak of 128 *zapovedniki* slid to 79 by 1968, with the total area roughly halved (from 12.5 to 6.4 million hectares). However, in 1977 nature protection was inscribed in the constitution (Obertreis 2018), giving a more direct legal basis for citizens to call the government to account. And so they often did, even before the enactment of such laws. This came about especially by way of scientific communities in the biological and earth sciences, relatively independent outfits like the All-Russia Society for the Conservation of Nature and organising and consciousness raising by activists. Their efforts, facilitated by the ease of land use reallocation thanks to state ownership, are one reason for the eventual restoration, expansion and more effective protection of the *zapovedniki*. These successes in turning the anti-ecological tide of the 1960s were also due to the actions of sympathetic forces within government and the wider scientific establishment (Pryde 1972, 20–1; Weiner 1999).

Conservation was not limited to *zapovedniki*. In 1983 the Andropov government set up tens of national parks for alternative, recreational uses (Ostergren and Jacques 2002). In the main, national parks served environmental education and cultural heritage purposes as well as providing additional areas for ecological monitoring. These were complemented by millions of hectares of natural refuges (*zakazniki*), characterised by variable uses and hunting restrictions or prohibition. Lastly, natural monuments (*pahmiatnyki prirody*) were limited to protecting certain landscape features, which could range from an endemic plant to an entire coastline. Such conservation areas preceded and/or were expanded or newly formed during the USSR's existence. By the time the USSR was dismantled, cutting-edge ecological work and protected representative ecosystems were contributing to improving prospects towards ecological sustainability. Even as this chance was largely sacrificed to capitalist prerogatives by the 1990s, the size and degree of protection of these nature reserves remain historically unparalleled and is among the ways in which a high degree of biodiversity was successfully preserved in the USSR (Colwell et al. 1997, 58).

Soil and Forest Conservation in the USSR

Following in the footsteps of a fledgling movement inspired by the ideas of V.V. Dokuchaev, the eminent late nineteenth-century geographer and founder of soil science, soil scientists finally got the chance to set up field experimental stations to address soil erosion after the October Revolution. The first one dedicated to soil conservation studies was put together in 1921 and eventually seven were established, backed up by two national soil conservation institutes: one for farming issues, the other for the rest.

The 1948 'Plan for the Transformation of Nature' was a major boost for soil conservation. Farming policy mandated afforestation for the first time, along with the expansion or overhaul of irrigation systems. Forest shelterbelts complemented reservoir construction in more than 120 million hectares to reduce the impact of seasonal soil-erosive hot dry winds. The Dnepr, Don, Ural and Volga catchments were thereby blanketed with protective woods and other vegetation. The largest planting effort, in 1949–53, succeeded in securing windbreaks for more than two million hectares. In this manner much of the Russian Plain's soil surface, mainly in the steppe and forest-steppe zones, was stabilised and protected for the long term. The increasing rates of erosion affecting such soils had been a topic of concern since the 1600s, but little had been done about the problem until the Bolsheviks came to power. Among the beneficial results of the policy is that the upper humus-rich soil layers (topsoils) of chernozem soils have become thicker over time under windbreaks compared to chernozems untouched by major human impacts like farming. Chernozems (grassland soils) are prevalent in the steppe regions. Their stabilisation and topsoil thickening constitute an especially important feat because they are often more erodible (due to the frequent loess sediment substrate). In addition, promoting chernozem topsoil development increases long-term carbon sequestration in soil organic matter. Unfortunately forested windbreaks have increasingly been destroyed since the early 1990s (Chendev et al. 2015).

Timber demand from industry and for export also put pressure on forests, in continuity with Czarist times, but the logging was much more sustainable overall under Bolshevik rule. This was because timber extraction was heavily tempered by the afforestation and reforestation programmes introduced in the 1950s, as discussed above. By the late 1960s most timber was exported from the USSR to countries like the UK and Japan. Of total timber exports, a fifth went to the former and a third to the latter (Goldman 1972, 167–8). Deforestation and ensuing soil erosion, including near Lake Baikal, was not traceable to socialist state mismanagement. According to Goldman (1972, 169), much deforestation occurred even earlier, in spite of forest protection policies under Lenin and later Stalin, and because of similar economic pressures (Brain 2016). Yet total forested area consistently rose from 738 to 792 million hectares between 1961 and 1982 (Barr 1988, 249). In the Western and most industrialised part of the USSR (west of the Urals), the extent of forest cover reached more than 216 million hectares by 1985. The afforestation and reforestation policy set a pattern that outlasted the USSR, so that by 2012 the total area was expanded by another ten million hectares (Potapov et al. 2015). Such historical and current forest expansions have mitigated the effects of clear-cutting in Siberia since the early 1990s, which had not happened in the USSR to anywhere near the more recent levels of devastation (Achard et al. 2006; Trunov 2017).

As the USSR government engaged in rapid industrialisation, agricultural area was expanded. Pressure on forest ecosystems was counterbalanced by shelterbelt construction, but, with increasing mechanisation, farming eventually came to be more impacting on soils and, to some extent, laden with agrochemicals. Especially in the Russian Plain, the expansion of beet production alongside cereal crops was accompanied by deeper ploughing and heavy machinery, causing compaction problems that were not adequately addressed until the 1980s (Chendev et al. 2015). Soil compaction often augments soil erosion rates. However, in the forest-steppe zone of Kursk oblast, for instance, a long-term study on variously managed and uncultivated chernozems (1964–2002) found that even the worst farming machinery effects of the 1970s left the soil's physical characteristics relatively undamaged for such soil types (Kuznetsova 2012). Compaction-level measurements, like soil bulk density and porosity, were within permissible levels (1.25–1.35 g cm^{-3} and 45–55 per cent respectively).

A major turn was the so-called 'Virgin and Idle Lands Programme' (VILP). Between 1954 and 1964 grassland cultivation was extended to northern Kazakhstan and western Siberia. There was ample precedent though. The 1941 Nazi and fascist invasion and extensive farmland destruction had already prompted an eastward shift in farming area to include Kazakhstan. By 1951, 2.6 million hectares of land were put to the plough (McCauley 1976, 167–8).

The areas affected by the VILP are characterised by wide-ranging interannual precipitation levels, dry and high spring/summer winds (90–110 kph) and recurring droughts. A primary incentive for the VILP was, aside from generating more revenue for industry and military power development, a need to raise grain production and augment meat and dairy availability. For example, growing wheat on the steppes would allow fodder crop specialisation in other regions. To a major degree the objective was to sustain an increasingly urban and wage-dependent population.

Eventually 23 million hectares of grassland came under the plough with the help of 650,000 colonists, despite specialists' reservations based on the experiences of major crop failures in the 1930s. Worse, the VILP featured an anti-fallow campaign, eradication of pre-existing vegetation (especially with mouldboard ploughs), machinery-induced compaction and constant pressures on state and cooperative farms to maximise yields and forsake field rotation, contour ploughing and other soil protection techniques. Matters

were little helped by limited irrigation, or by agrochemical distribution and application difficulties. The combination of monocultural
production with insufficient herbicides created a bounty of undesirable plants (weeds) eventually contributing to reduced grain yields.

These considerations benefit from historical hindsight, however.
The Khrushchev government had legitimate reasons to be sanguine
about the VILP. In 1956 wheat production hit record levels and the
1958 harvest proved to be very generous. But excellent yields alter-
nated with poor harvests in tune with variable yearly weather patterns
(McCauley 1976, 186). Moreover, viewed over the spans of decades,
successes could be lethally deceptive. A prolonged drought (1961–3)
led to major productivity losses and more frequent dust storms.
Approximately seven million hectares of newly established cropland
had to be retired. The VILP also led to land concentration under
fewer, larger state farms, mostly at the expense of remaining pas-
toralists and cooperatives. The calamity turned the VILP into a net
financial loss for state coffers, spelling more dependency on North
American imports, and contributed to the removal of Khrushchev
by rival forces by 1964. The succeeding Brezhnev government dis-
continued the policy and instituted a special soil erosion programme
in the Ministry of Agriculture. Not that this could alter the cyclical
nature of meteorological conditions, but the new measures did
attenuate erosion problems.

Interestingly it was to Canada that the government looked for
models, not the US or Argentina, even if they arguably had similar
problems and grassland ecosystems (Josephson et al. 2013, 152).
However, in at least some portions of those very Canadian prairies, soil
erosion was far from under control at the time and in some respects
continued to be severe in the uplands (Gregorich and Anderson 1985;
Martz and de Jong 1987). It was not until the 1980s that conservation
tillage, crop residue management and other measures succeeded in
substantially reducing erosion rates (Huffman et al. 2000). It should
be emphasised that the knowledge transfer also went in the opposite
direction. Techniques developed in the 1930s in the USSR for industrial production systems to cope with permafrost conditions were
borrowed and adapted by North American firms and governments
(Chu 2018, 184).

More importantly, continued expansion of farmland subsequently contributed to reducing grain import dependence, as drought started to affect other regions. Noteworthy is the increasing aridification of the regional climate in Central Asia, with the increasing frequency of ever more intense droughts since the middle of the 1960s (Elie 2018) – Khrushchev's VILP was a case of awful timing. It must also be acknowledged that, unlike countries such as the US and Canada, the main farming areas of the USSR were located exactly where drought frequency and magnitude had been rising. There was no possibility for the USSR government to switch food production to another region. And unlike some major capitalist countries, the USSR did not have or seek colonies or land abroad under corporate control to overcome domestic agricultural limits.

The story is not too dissimilar from what occurred in the US with the Dust Bowl, but there are several crucial differences. One is that the USSR, unlike the US (or Canada), is prone to drought except in the north, where farming is largely unfeasible or impossible. Expanding farmland wherever it is practicable is a way to ensure that agricultural production is much less disrupted when one region is hit by drought (McCauley 1976, 196). In the US, unlike in the USSR, the drought-mitigating effects of irrigation systems, which were expanded rather than reduced, have robbed water from many Indigenous communities. They are also behind one of the world's biggest, yet seldom discussed, environmental disasters. The Ogallala aquifer in the western US, one of the largest in the world, is now threatened with depletion and, in some areas, salinisation (Brooks and Emel 1995; Woodhouse 2003). Something similarly disastrous may be in the offing with the aquifers in the region upstream from the Aral Sea (Gadaev and Yasakov 2012). Additionally, human-accentuated droughts continue to generate periodic dust storms to this day in the south-west US as they do in the steppe regions, including Ukraine (Birmili et al. 2008; Dudiak et al. 2020), and they are major driving and respiratory hazards.

In any event, following the Khrushchev administration matters did improve overall. Impacts on grassland soils were reduced under the Brezhnev government with the introduction and spread of minimum tillage, spatial alternation of annuals and perennials and re-establishment of grasses, among other soil conservation methods. Previous practices were also reinstated, such as crop rotation, contour ploughing and tree planting for shelterbelts, which reduce vulnerability to wind erosion (Elie 2018; Kraemer et al. 2015; McCauley 1976; Zonn et al. 1994).

These soil conservation policies were far from easy to introduce and spread. When the USSR became a net grain importer (mainly from the US) in 1972, a third less land started to be left fallow (unsown to regain fertility) as in previous times. In the official press, open criticism was levelled at the government during the 1970s and 1980s for failing to address soil erosion effectively, including by top soil scientists (Brown and Wolf 1984, 17–18).

Nevertheless, soil conservation measures implemented from the 1970s onwards were eventually largely successful. Consequently much of the soil in the Russian (especially the central portion) main area of cultivation is now stable, under grasses or woods, and mainly free of erosional gullies and ravines. Techniques varied by environment type and include: revegetation, gully infilling, forest belts, periodic fallow and planting perpendicular to slope direction (Brown and Wolf 1984; Golosov and Belyaev 2013). There were also moves to expand organic farming, which by 1980 had spread to cover 4 per cent of farmland (Peterson 1993, 109). Moreover, thanks to the recentralisation of forestry away from the Sovnarkhozy (Regional Economic Council) and under relevant ministries, deforestation was also held in check (Josephson et al. 2013, 162). Farmland and logging area abandonment since 1990 may have helped restrain or halt erosion rates (this depends on the degree to which human impact reduced erosion rates relative to prevailing or changing environmental conditions), but it is the net positive result from USSR policies that have made the difference, as shown in comparison to the higher erosion rates in adjacent countries (Sidorchuk and Golosov 2003; Wuepper et al. 2020). One could speak of a positive USSR legacy in the case of soil protection measures. If the status of soils were as disastrous as portrayed by mainstream scholars, foreign firms and government would not now be flocking to countries in the former USSR to buy up farmland (Visser and Spoor 2011).

Pollution Abatement in the USSR

Major pollution problems accompanied rapid industrialisation and decades of increasing industrial output. A much justified and oft-cited complaint or critique about the USSR was the lax or lack of enforcement of what were otherwise much more stringent water and air pollution standards compared to contemporary ones in the US (Goldman 1972, 27–8; Oldfield 2005, 93–4). However, it is also true that legislation to reduce air pollution introduced under Czarism in 1913 was never enforced until the Bolshevik Party took over and eventually expanded the degree of restriction (Pryde 1972, 12).

Water pollution remained a problem throughout the latter part of the USSR period, but there were also many cases of successful prevention and remediation (Pryde 1972, 144). Under the Stalin government, public health infrastructure and measures against air pollution were introduced (Josephson et al. 2013, 87). Major investments were made in pollution prevention research and implementation by 1929–30, but the wartime invasion and devastation interrupted and debilitated the planned infrastructure and applications over the long term. Nevertheless, by 1949 all industries and building projects were required to install dust-trapping devices (Izmerov 1973). It was during that administration as well that, in 1951, some of the world's first maximum permissible concentration standards were established, at first for 122 pollutants. These scientific breakthroughs laid the foundation for toxicological standards worldwide (Izmerov 1973, 28, 45; Josephson et al. 2013, 92). Air pollution legislation was updated every five to ten years, and by 1970 comprehensive national guidelines on water use and pollution prevention were introduced (Pryde 1972, 137).

By the 1970s regular air quality monitoring was spread over 110 cities, reaching 150,000 total point and transect samples (Izmerov 1973, 35). In the 1980s, 537 cities were equipped with such monitoring infrastructure, totalling 1,155 air quality monitoring stations (Permitin and Tikunov 1992; Pryde 1991, 32). Additionally, tens of background monitoring stations were built to reach complete country coverage by the early 1980s. All major contaminants were included, in addition to several heavy metals like lead and cadmium and pesticides like DDT. These stations, following a decade of disruption and retrenchment, form the backbone of pollution monitoring today in the former USSR countries (Chernogaeva et al. 2009). Thanks to copious scientific research and monitoring over decades, the sources, diffusion routes and health effects of air pollution were identified, helping to make the design and implementation of plans and regulations more effective in tackling the environmental challenges of industrialisation.

As a consequence, alternative technologies for heating, electricity and coal combustion were developed and applied to reduce incomplete combustion, sulphur content and total emissions. This included switching to natural gas and hydropower sources, as well as using less polluting forms of coal. Fuel use based on natural gas increased roughly tenfold from 1955 to 1987, from 2.4 to 25 per cent of total energy sources (Shahgedanova and Burt 1994). The result was a marked diminution in airborne dust and sulphur in urban centres within a decade. For example, sulphur levels were more than halved and ash amounts reduced by two-thirds between 1953 and 1961 in Moscow. Similar outcomes were achieved in many other towns and the positive trend continued through to the USSR's disappearance (Izmerov 1973, 63–4, 104–5; Shahgedanova and Burt 1994, 208).

Pollution mainly from the metallurgical, petrochemical and power industries remained unresolved (Izmerov 1973, 65). Heavy metals contamination, for example, became severe in some areas by the 1970s, but the case of lead shows that there were also countervailing tendencies that rivalled the wealthiest capitalist countries. Leaded paint was already banned by the 1920s. In the US this did not happen until 1971. The use of leaded petrol, which did not reach the USSR until the 1940s, was severely restricted within many cities and tourist destinations by 1956 (Thomas and Orlova 2001). These are major reasons behind the much lower incidence of lead contamination, but stationary sources like the lead smelter in the Rudnaya River Valley (Russian far east) were highly and enduringly contaminating of proximate environments and people's bodies (Kachur et al. 2003).

Eventually, after peaking in the early 1980s, sulphur emissions from those stationary sources were successfully cut overall and drastically so, from 118 to 30 μg m^{-3}, with localised exceptions. Nitrogen oxides and carbon monoxide emissions, on the other hand, were largely held in check, mostly because of investments in public transport and the mitigation of private motorised vehicle use (Shahgedanova and Burt 1994, 212, 216, 219). There were nevertheless substantive reductions in industrial emissions by the middle of the 1980s. For example, pollution within a 30 km radius of the Middle Ural copper smelter showed rapid diminution of total emissions from 225 to less than 175 tonnes per year within the span of a couple of years in the early 1980s and then steadily declining further at lower rates. This includes sulphur, dust, nitrogen oxides, arsenic, copper and lead emissions. Upper soil horizons, though, retain long-term toxicity from decades of heavy metal and metalloid deposition (Vorobeichik and Kaigorodova 2017).

Though air pollution never reached the levels of countries with comparable industrialisation levels, like Australia, the US and Canada (Hill 1997, 20; Oldfield 2005, 24–5), major reductions in sulphur and nitrous oxide emission were achieved thanks in part to a switch to natural gas (Figures 4.1 and 4.2). In this manner, particulate matter was also reduced over time (Figure 4.3), doing better than the US in most years. Most of all, better results were thanks to 'an effort initiated by the government in the 1970s to improve air quality' by using multiple mitigation techniques in stationary sources, such as retrofitting with more effective scrubbers (Peterson 1993, 45). A negative effect was the rising methane emissions, in part due to the fuel switch policy, but unevenly over time and surpassing US emission rates (Figure 4.4). However, CO_2 emissions were always considerably lower than those in the US (Figure 4.5). The USSR may have had the highest sulphur emissions among the industrialised state-socialist countries (also because of reliance on lignite), yet it was always below US emission rates and also the most affected by long-range air pollutants from countries like the UK, Denmark and West Germany, as well as Poland (Hill 1997, 208; Josephson et al. 2013, 220–1; Pryde 1972, 152).

Water pollution was mainly a locality-specific problem primarily associated with small streams in proximity to industrial plants and mines, as recognised at least by the late 1930s. The introduction and diffusion of sewage systems and treatment plants, along with increas-

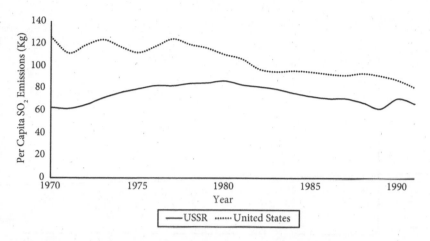

Figure 4.1   Per capita SO_2 emissions (kg) in the USSR and US, 1970–91
Sources: Data from EC-JRC (2020) are divided by country population data from World Bank (2021b).

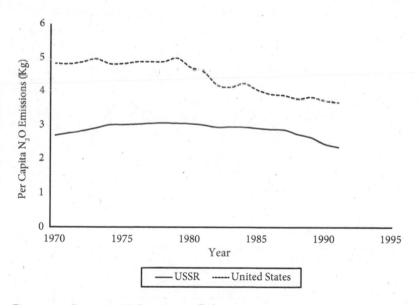

Figure 4.2   Per capita N_2O emissions (kg), 1970–91
Sources: Data from EC-JRC (2020) are divided by country population data from World Bank (2021b).

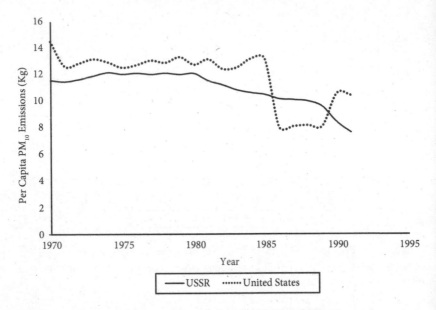

Figure 4.3 Per capita PM$_{10}$ emissions (kg) in the USSR and US, 1970–91
Sources: Data from EC-JRC (2020) are divided by country population data from World Bank (2021b).

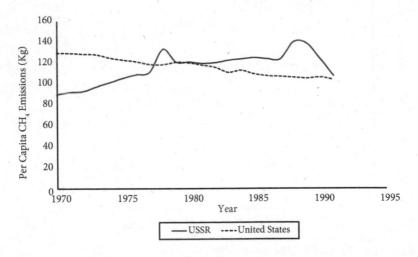

Figure 4.4 Per capita CH$_4$ emissions (kg) in the USSR and US, 1970–91
Sources: Data from EC-JRC (2020) are divided by country population data from World Bank (2021b).

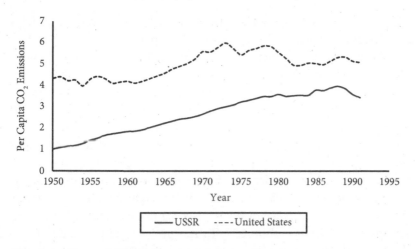

Figure 4.5 Per capita CO_2 emissions (tonnes), 1950–91

Source: Boden and Andres (2014).

ingly stricter mandates for industries to ensure water purification prior to emission, aided in preventing or mitigating pollution in most cases. With postwar reconstruction and massive expansion of drinking water and sewerage infrastructure, more than three-quarters of USSR inhabitants had access to piped, potable water by the end of the 1950s (Litvinov 1962). Along most of the stretches of the larger streams, such as the Volga and Yenisey, dissolved oxygen levels were adequate or good from an ecological standpoint (within and higher than the recommended 6.5–8.0 mg L^{-1}). While nitrate levels tended to be low, ammonia concentrations were above recommended 0.5 mgL^{-1} by 1980 in the Ob and Amur rivers. By 1985 concentrations were successfully reduced to below recommended maxima (Zecchini 1996, 53). However, by the 1980s the Don and Dniestr were highly polluted from a combination of industrial, municipal and agricultural effluents (Oldfield 2005, 25).

Part of the reason for the reduction in total toxic emissions over time was due to increasingly stringent regulations, actions to shift fuels, improvements in technical efficiency and strides in reuse and recycling of waste in industrial processes (Dudenkov 1985; Pryde 1972). More fines were introduced for enterprises caught polluting, even if enforcement had a mixed record. Some interpret the policies as ineffective because fines were lower compared to the loss of funds for failure to meet production targets (Josephson et al. 2013, 198–9). Most trace this to an intrinsic conflict of interest (e.g. Josephson et al. 2013, 203; Goldman 1972; Pryde 1972), as if a state is an internally coherent or homogeneous unit, a monolith impermeable to the rest of society (see Oldfield 2005). To think about the matter this way leaves unexplained the cases in which state institutions fought each other over conservation issues or resource management and the instances where conservation prevailed over economistic calculation, including by means of environmental campaigns from below (Weiner 1999). Such arguments also cannot explain government actions to replace fuel sources with less polluting ones, as described above. As further examples, in the 1960s pollution-abating retrofitting was installed in thousands of industrial establishments, and a hundred polluting factories in Moscow were relocated away from expanding residential areas. In the 1970s, 35,000 propane-fuelled lorries and public transport vehicles were ordered for Moscow to reduce carbon monoxide emissions by three-quarters. By 1972 central heating district facilities reached most domiciles in the largest cities and reduced pollution from coal use (McIntyre and Thornton 1978). By the 1980s, thanks to both public pressure and concerted industrial policy efforts, pollution levels were coming under control and falling (Oldfield 2005, 24). In large measure the limited successes in fighting industrial pollution had to do with the partial defeat of environmentalists and public health practitioners at the hands of the economic ministries (Kelley 1976; Ziegler 1980).

All those environmental struggles within the USSR are effaced by an overwhelming emphasis on lasting pollution brought about through nuclear power. The first is associated with an explosion at Kyshtym in 1957, at the Mayak plutonium production plant in Chelyabinsk oblast (Josephson et al. 2013, 132; Pryde 1972), making it the third biggest nuclear accident behind Fukushima Daiichi (2011), Japan, and Chernobyl (1986). Less publicised these days is the Windscale fire at the Sellafield nuclear facility (UK), which happened less than a month after the Kyshtym accident. Ranking among the top ten nuclear disasters, the Windscale accident released radioactive fallout through the UK and parts of the rest of Europe. The largest such accident was, of course, Chernobyl. So much has already been written about it that the few remarks included here are rather to supplement what was discussed in Chapter 3.

The siting of Chernobyl made for a much greater impact on land, rather than mainly the ocean, as in the case of the 2011 Fukushima nuclear meltdown (an ongoing disaster at the time of writing; Kasar et al. 2020). Radioactive releases largely escaped westward from the nuclear accident site, and most fell on communities in what is currently Belarus. The number of evacuated people was more than twice that of Fukushima, in part because of physiography and prevailing wind patterns, but mainly because of the USSR authorities' decision to locate the reactors in a region with a relatively high population. The problem is not whether the Lenin nuclear plant meltdown near Chernobyl (Ukraine) was or was not as disastrous or deadly as made out to be. A disaster it was and mortality rates are likely high in the long aftermath. The problem is that Chernobyl is seared in public memory of the 'free world' as proof that socialism leads inexorably to catastrophe. Chernobyl is almost like a stock image archived permanently under a folder labelled 'communism'.

Yet the disaster and its pictorial representations are a double-edged metonym. As much as it is a way of compelling us to understand communism as intrinsically horrific, Chernobyl is also a threat to the nuclear industry and to all those environmentalists (leftist or not) who view the nuclear option as a viable alternative to fossil fuels. Protestation must never exceed the bounds of moderation towards the nuclear industry. The nuclear disaster must be understood as only demonstrating the incompetence on the part of the USSR. We are supposed to think that such a catastrophe will never happen under technologically superior free-market democracies, or under fully automated luxury communism, according to some on another part of the political spectrum. In none of those social systems would there be any use of graphite-moderated BRMKs (high-power channel-type reactors), as in Chernobyl. Never mind that there are nine BRMKs still in operation and without any particularly greater environmental destructiveness than usual for nuclear power.

What the Chernobyl accident also demonstrated was a convergence of interests among top-ranked USSR and US officials and the nuclear industry. Focusing on the accident as a uniquely destructive event takes attention away from the destructiveness of nuclear power in general. As Brown has pointed out:

> In four decades of operation, the Hanford plutonium plant near Richland [Washington State] and the Maiak plant next to Ozersk each issued at least 200 million curies of radioactivity – twice what Chernobyl emitted – into the surrounding environment. The plants left behind hundreds of square miles of uninhabitable territory, contaminated rivers, soiled fields and forests, and thousands of people claiming to be sick from the plants' radioactive effluence. (Brown 2013, 3)

And there were and remain tens of those plutonium production centres. The Hanford site in eastern Washington (US) is more than just a production and disposal operation. In 1949 there was a deliberate release of 7,000 to 12,000 curies of radioactive iodine as an experiment in monitoring radioactivity levels. The authorities compromised and cut short the lives of thousands of downwind inhabitants, predominantly in Colville, Coeur d'Alene, Kalispel, Kooten-ai, Nez Perce, Spokane, Umatilla, Warm Springs and Yakama Indigenous communities. To this day there is no adequate epidemiological study.[2] Chernobyl was accidental; Hanford was intentional.

This in no way absolves or justifies the USSR government's nuclear energy programme. It may have been expedient in raising energy provisioning and moving away from more polluting sources and there may be still more reasons that the USSR felt compelled to engage in a nuclear power programme, but the disastrous consequences overwhelm the benefits with or without accidents. Any socialist movement, especially when attaining political predominance, should learn from the state-socialist nuclear power programme so as to stay clear of any such technology as well as devise safe ways to decommission and dispose of spent nuclear fuel.

---

2 https://hibakusha-worldwide.org/en/locations/hanford.

Producing energy from nuclear fusion, however, is a much more promising technical alternative. Just as in the case of developing hydrogen-powered vehicles (Fateev et al. 2020; Korovin 1994), the USSR was at the forefront of nuclear fusion research. In 1951 nuclear physicists and engineers there began to develop a device containing hot plasma by using magnetic fields and that became the precursor to the Tokamak (Smirnov 2009). The Tokamak is the current basis for developing a nuclear fusion reactor, now taken to the most promising levels by scientists in South Korea (project KSTAR). It remains to be seen whether, just as in the case of solar panels and electric cars, such a technology is put in place through socially oppressive means and creates other kinds of environmental destruction.

The USSR's Impacts on Arctic and Aral Sea Environments

There is just as mixed a record relative to parts of the Arctic and the Aral Sea ecosystems. In the Arctic zone, the Kola Peninsula in particular was hard hit from the 1930s onwards with the introduction and spread of factories, mines and eventually nuclear power, leading to a corrosive cocktail of acid rain, deforestation, polluted waters, upturned earth and, initially, prison labour, in continuity with the Czarist dictatorship (Bruno 2016, 26–7). However, when, for example, the Stalin administration turned their attention to the region in the 1930s to mine apatite, they had not set out to wreck the environment. There were careful plans to locate mining areas well away from urban centres and ecologically sensitive areas and to maximise the reuse of mined materials and spoils (such as nepheline for aluminium), but discovery of bauxite in the Urals made the mineral recycling programme redundant and, as the Urals' deposits dwindled in the 1970s, importing bauxite was cheaper than recycling mine waste from the Khibiny Mountains (Bruno 2018). It was not the Stalin regime that was careless about the environment, in this instance. The greater integration into the capitalist world economy by the 1960s, with the Khrushchev and Brezhnev administrations, created the conditions whereby economic pressures increased the influence of a commercial valuation process, where displacement of environmental destruction is transformed into cheaper raw material import. This intensified an already highly problematic set of resource extraction policies to aid rapid industrialisation and that resulted in much social and environmental harm.

Along with the largely deleterious impact on many species, such as reindeer (partly herded into large farms), USSR policy, including forced sedentarisation, displaced and marginalised the Sápmi, Komi and Nenets. The Sápmi suffered similarly just across the border, under social democracy. As environmental historian Andy Bruno has observed, 'Soviet rule had turned the Kola Peninsula into the most populated, industrialized, and militarized section of the global Arctic, as well as one of the most polluted' (Bruno 2016, 4). None of this was preordained and things could have been turned about in more positive directions. Nickel smelters, among the more polluting industries, turned into a veritable scourge mainly when the USSR came under more intense economic pressures in the 1970s. Agricultural output shortfalls and accompanying dependence on US grain imports, and then the 1980s major oil export revenue losses due to a world oil glut, combined to intensify indebtedness to Western capitalist institutions (Bruno 2016, 19–20). *Zapovedniki*, nevertheless, were established (e.g. the Lapland preserve) and were valiantly defended by dissenting communist conservationists even at the height of the 1930s purges and executions and despite multiple attempts at opening *zapovedniki* up to industry (Weiner 1999).

Then there is the Aral Sea, shrunk as a direct consequence of the diversion of major rivers to suit hydroelectrical purposes and to provide irrigation privileging cotton production as an export-oriented cash crop. As already remarked, it was under the Czarist regime that major inroads were made towards turning that part of Central Asia into a cotton-producing region. It was not until 1945 that substantive river redirection, mainly the Syr and Amu Darya, started. Canal construction proceeded through the 1970s and extensive soil salinisation problems eventually followed. By the 1960s the sea was expected by experts within the USSR to turn into a salt marsh by 2000 (Goldman 1972, 216–17). More recently the sea was predicted to have disappeared by 2021 (Gadaev and Yasakov 2012, 10). Fortunately this has not turned out to be the case, but it is small consolation that a mere fraction of the sea remains and requires massive investment just to retain its current, much diminished size. The impact is not just on the Aral Sea and its communities and ecosystems, particularly the now mostly disappeared wetlands. There is a regional effect in making an already arid region even drier and with greater seasonal temperature extremes, as the moderating effect of a large body of water are largely gone (Laity 2008, 282–3). The impacts were not entirely negative either. Efforts at mitigating vegetation losses and reforestation succeeded and stabilised surrounding ecosystems (Kariyeva and van Leeuwen 2012; Robinson 2016; Zhou et al. 2015).

Yet the sea's shrinkage was not due to the USSR alone. Closer scrutiny reveals that the most extensive shrinkage of the sea happened after, not during, the USSR's existence. This is evinced in 1977–2010 NASA satellite imagery, which shows that it was after 1990 that the sea was reduced to a quarter of its 1960 extent (NASA 2012). Most of the losses occurred after 1989, even as there were temporary declines in water withdrawals and increases in precipitation and upstream runoff (Micklin 2007, Wang et al. 2020). This is in part due to open and unlined irrigation canals as well as the continuation of cotton production combined with another water-demanding cropping system, wet rice monocultures (Laity 2008, 282). In addition, temperatures had been markedly increasing decades prior to the dissolution of the USSR (Lioubimtseva 2014), magnifying evaporation rates and speeding up the sea's recession. The process was at first gradual, over the 1960–80 period, which is part of the reason that USSR governments underestimated the problem. It took a few decades for the shrinking trend to become prominent (Glantz 2007) and for regional climate warming to exacerbate the impacts of river diversion. The warming trends, as shown in Chapter 3, are hardly the making of the USSR alone or of socialist states more broadly. The Aral Sea disaster is regrettably not unique. Countries like the US are even more wasteful with water, with only a third to half of irrigation water reaching crops. Many examples exist in the US and elsewhere of shrinking lakes tied to industrial farming schemes in semi-arid and arid regions (Horowitz 2012; Pryde 1972, 123).

Nevertheless, decisions were made in full awareness of the disastrous consequences. Revenues from cotton exports were deemed more important. Downstream fisheries were already deleteriously impacted by the 1970s. Within a few decades 500,000 hectares of spawning and migratory areas succumbed not only to desiccation but also to the poisonous cocktail of fertilisers and biocides from industrialised farming. Still, with central planning coordination, including the introduction of aquaculture, fishing remained viable and production even increased (Glantz 2007; Karimov 2011; Petr et al. 2004). Local people's health was neglected and thousands of lives were shortened by the toxic consequences of polluted air and water, as well as threatened by the leftovers of a biological weapons programme (Edelstein 2012). Cotton production soared, doubling between 1960 and 1980, contributing substantially to the state acquiring core capitalist currency to repay loans and disproportionately high interest, and, among other things, purchase high-technology products. This, more than any notion of mismanagement or inferiority of central planning, can explain the ecologically and, for part of the region, socially destructive outcome, given that endless capital accumulation was not the overall objective. It was certainly not some reckless intervention in the environment intrinsic to 'socialism', as is often portrayed.

Contrast this with the main goals of outfits like the US Army Corps of Engineers over the same period or more recently. This is an analogy made by some regarding similarities in the environmental practices of the USSR and the US (Josephson et al. 2013, 119), except that the objectives were rather different. Among other things, the aim in the US was to improve river navigation and build dams as part of infrastructure primarily serving commercial interests or, in cases like New Orleans, mainly the needs of upper-class whites. In the meantime, the projects undertaken by the Corps often led to destroying wetlands and degrading riverbeds, among other deleterious consequences (Hartman and Squires 2006; Robinson 1989; Schneiders 1996). The dams on the Colorado River and other diversions of river water in the western US have had regionally destructive repercussions to benefit mainly white settler farming businesses and at the expense of Indigenous communities' access to water, not to mention the submergence of sacred places and former settlements (Young 1997). Downstream of the Colorado River dams there are problems of salinisation and toxic chemical build-ups in the river delta, while in Mexico much of the wetlands area of Baja California has been destroyed. In fact, the Colorado River does not reach Baja California anymore, thereby starving local marine life of nutrients and reducing biodiversity (Kowalewski et al. 2000; Tecle 2017). One could claim that the intent was not necessarily destructive and that if there was destruction it is not intrinsic to liberal democracy; carry the same argument over to state socialism and we get the same degree of absolution. The problem, though, is the thirst for endless capital accumulation, the basis of liberal democracy that eventually drives even the most socially or ecologically positive intentions towards awful outcomes.

The USSR's Environmentally Constructive Contributions

Even if the USSR's impacts were at times destructive, and on a couple of occasions lastingly catastrophic, policies to contain environmental impact and promote conservation were much more successful than typically acknowledged. Ecological sensibility was diffused to tens of thousands of people through scientific and volunteer association. Conservation research enjoyed high levels of investment and infrastructure (Pryde 1972, 164). This all made it easier to set up ecological preserves, establish biodiversity conservation areas and organise to pressure government to improve air and water quality and to fight off some large-scale plans of environmentally destructive potential. Centralised decision-making processes and the drastic curtailment of private land ownership made conservation more effective, aiding in the set-up and protection of large preserves throughout the country. This made possible the institution of the world's most comprehensive system of ecological preserves (*zapovedniki*), which was enlarged over time, aiding in the protection of biodiversity and restoration of threatened species. Major strides were made in soil conservation and afforestation whose positive effects are felt even today. Pollution problems existed with industrialisation and increased in severity especially in the 1970s, but technological improvements, fuel source switching to methane, increasing restrictions and fines, as well as grassroots activism, contributed to substantial and steady reductions in emissions by the 1980s.

There were other important developments that would require a separate book to recount. Zoning and industry siting were more easily controlled with centralisation, with potential for ecological and social benefits contingent on the outcomes of struggles within and between state organs at different levels as well as with organised locals. The suppression of hydroelectric dam construction in Armenia and a paper pulp plant near Lake Baikal exemplify successful struggles.

Urban centres were designed to limit excessive consumption and waste and promote socialisation. Heating and hot water systems were made more energy efficient and available to most citizens, thanks in part to urban planning that focused more on providing communal living spaces. Amenities reached most everyone, and not just in cities. There was a thick network of public transport, plenty of urban green space and very little private vehicle use. Consequently air pollution was largely confined to stationary sources like factories, which were usually located well away from inhabited areas, and green spaces helped absorb contaminants. Workplaces were mainly located close to people's residences or reachable by public transport, pre-empting the need for personal motorised vehicles for most. By 1970, with an eye to reducing fossil fuel use further, electric cars were being developed and tested (Korovin 1994). This would have complemented existing reliance on railways for mass freight of consumer and producer supplies, which contributed to keeping mobile sources of pollution low (Krausmann et al. 2016).

The lack of mass consumerism prevented the existence of premeditated product obsolescence. An advantage of suppressing commercialisation and greater media control was the existence of little to no advertising, for example. There was no 'unrestrained competition to seduce the consumer into buying what he doesn't need by the use of gaudy and unnecessary packaging, misleading advertising, constant design changes, psychological pressures, "beautifying" chemical additives of uncertain side-effects, etc.' (Pryde 1972, 164). A result was 195 kg of solid waste per capita for a total of 56 million tonnes in 1988, compared to 665 kg per capita in the US and 162.9 million tonnes total (Peterson 1993, 130). There were few to no disposable products and thereby much less solid waste accumulation. Recycling rates were widely promoted and much greater in the USSR than in the US (Goldman 1972, 174). Compared to the US and other high-consumption countries, the USSR was also more advanced in the recycling of human waste, including using biosolids for farming (Goldman 1972, 273–83). The latter, it should be noted, can create pollution problems that at the time were not adequately studied or not considered at all. With proper screening and processing, however, recycling effluent can and should be pursued further – much further than has been the case since the 1970s.

Lastly, there were important and lasting positive environmental legacies of the USSR at the global level by way of international environmental agreements and treaties, along with the copious scientific work and data needed to back them up (Foster 2015). Starting in the 1950s, climatologists and geophysicists in the USSR developed tools to understand the global climate system that would become essential to climate change study. By the 1960s they were analysing the role of human causation in global climate change and developing climate change models grounded in palaeoclimatic analogues to enable forecasting. They were internationally influential participants from the late 1950s in the precursors to the 1990 Intergovernmental Panel on Climate Change, where USSR scientists were ultimately marginalised. Such precursors include the 1957–8 International Geophysical Year and the 1979 World Climate Conference, which was pushed by the World Meteorological Organisation under the leadership of Swiss and USSR scientists. The basis for multiple attempts at climate change protocols and agreements (so far complete failures) bears the heavy imprint of the USSR's precocious advocacy (Oldfield 2018).

The USSR, as part of the UN's Security Council, lent their political weight in support of the 1972 Stockholm Declaration, which outlined the principles of social and ecological sustainability. In 1975, partly owing to being a recipient of air pollution and acid rain from countries to the west, the USSR led the charge in pushing for an international treaty on long-range air pollution through the United Nations (Sokolovsky 2004). Negotiations were arduous and the USSR devoted major efforts, enlisting the backing of Scandinavian countries, to overcome four years of stalling and objections by NATO countries even to meet on the issues. The resulting 1979 Convention on Long-range Transboundary Air Pollution was a pivotal moment in the making of legally binding global environmental protection policies. Fostering scientific and technical collaboration among tens of countries to reduce air pollution, the agreement eventually proved successful in markedly reducing regional emissions in Europe and North America. Regrettably, the convention stands as a rare example of effective international action on the environment. The USSR also spearheaded the 1977 UN Conference on Desertification in Nairobi, with USSR scientists featuring as major organisers and ensuring that social causation would be among the main processes considered (Elie 2015).

As shown above, there were beneficial developments and promising potentials stemming from the institutional structures of the USSR. Ecological footprint data largely verify this both in total and per capita values. This becomes especially evident when compared to the consistently worse record for the US, with a similar population size, an ability to offload environmental harms on many other countries and a highly industrialised and much wealthier economy (Figure 4.6). Much could therefore be learned of ecological value from historical experiences in the USSR instead of just jettisoning them.

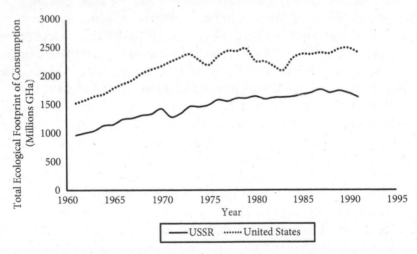

Figure 4.6   Per capita EFC in the USSR relative to the US, 1961–91
Source: Global Footprint Network (2019).

Among others, one is that making the economic much more evidently political is insufficient to tackle or prevent destructive environmental impacts without putting ecological issues front and centre of policy and everyday practices, especially by dedicating just as much if not more resources and rewards to ecologically constructive activities.

Another is that official pronouncements and legal frameworks must be aligned with material conditions and feasible prospects. The Communist Party leadership too often exaggerated what could be achieved under prevailing conditions. Furthermore, they often dissimulated the negative forms of environmental impacts as well as existing power relations, characterised mainly by worker subordination to central committee and multiple-level administrative and managerial dictates. This would have also meant avoiding claims of state-socialist superiority and dropping any visions of catching up to and surpassing the wealthiest capitalist countries in economic output.

Instead, it could have been easily demonstrated how much better off people were compared to the Czarist period and to almost all other countries in the world, in terms of living standards, for example. The realisation of the wealthiest capitalist countries' level of wealth relies on a history and a present of colonial and neocolonial horrors unbecoming of any socialist project. Rhetorical manoeuvres may have served a galvanising purpose for party members and maybe also wider society, but they eventually (if not immediately) backfired when environmental and social harms patently contradicted official statements and intentions.

Nonetheless, it is impressive that much environmental legislation and many positive environmental practices started to be instituted since the very establishment of the USSR, during protracted wars, famines and multiple scorched-earth foreign invasions. Conservationist rationales behind those policies, as discussed above, may have overlapped somewhat with those in countries like the US (e.g. securing a reliable internal supply of raw materials; see also Gare 1993), but the implementation of those policies faced much rougher economic circumstances from within and without. That so many environmental benefits were achieved in a rapidly industrialising country under such general strains testifies to the usefulness of the socialist state in mitigating environmental harm while raising standards of living in what was initially an overwhelmingly deprived agrarian society. The conflicts between environmental concerns and industrial interests, expressed institutionally through, for instance, the tension between pollution fines and production targets, signifies a symptom of the contradiction that was the USSR. It is no picnic to balance the striving to overcome capitalist and semi-feudal Czarist legacies of widespread material deprivation, grossly unequal access to resources and much environmental destruction with building socialism and ensuring a more liveable environment and the integrity of ecosystems.

THE PRC: FULCRUM OF WORLD ECOSOCIALIST STRUGGLES

Much maligned now from both left and right political positions, the PRC presents an interesting analytical and immediately political conundrum. On the one hand, if one recognises the PRC as socialist even after 1978, the environmental devastation that took off since then becomes evidence for the presumed ecological evils of state socialism. But this enmeshes all capitalist countries at once in such environmental destruction because of the tight connection between economic changes towards market mechanisms in the PRC and capitalist investments from and exports to the largest economies under liberal democracies like the US and Canada. On the other hand, if the PRC is deemed capitalist (and the official Marxist rhetoric spurious), then the obvious intensification and expansion of environmental damage since 1978 cannot be explained by socialist mismanagement at all. The destructive effects would be flatly capitalist in character, as in fact they are even according to the Cato Institute, a mouthpiece for the coarsest enthusiasts of capitalism. In their view, the redirection imposed by the Dèng government since 1978 led to a radical transformation of the PRC by the end of the 1990s, such that it 'is no longer communist except in name' (Coase and Wang 2013, 10). But if one takes environmental impact alone as the measure of ethics or progress, one could find praiseworthy the much lower impacts when the CPC was led by Máo. If anything, the PRC demonstrates most clearly what nonsense it is to hold the view that state socialism inevitably leads to environmental destruction. It is in this case just as arguable that it is the transformation of a socialist into a capitalist system that yields that kind of catastrophic trouble. More to the point, regardless of ideologically motivated labels, a decisive and massively more destructive turn is evident in China following the introduction of capitalist policies. This would appear to be the least controversial conclusion to draw so far and with far-reaching political consequences.

Legacies of Commercialisation and Colonialism in China

The ecosystems harmed or destroyed currently in China are often those that resulted from millennia of impacts from social systems preceding the 1949 establishment of the PRC. The highly heterogeneous ecosystems within the variably bounded territory of China, commercially oriented yet not capitalist since the 1600s or so (Marks 1998, 11–13), underwent major lasting and largely negative environmental impacts prior to the establishment of a socialist state. Among the most salient was biodiversity loss and deforestation associated with the expansion of an intensive form of farming whose aims were to produce surplus for trade and state, not necessarily to provide for subsistence needs. Matters were only exacerbated from the 1850s onwards with the belligerent encroachment of liberal democracies (the Opium Wars being among the more egregious examples), imposing terms of trade and commercial monopolies that essentially funnelled more resources away to foreign businesses. The forced subordination to a capitalist world economy dominated by European empires introduced or intensified plantation systems oriented towards export, concentrated in commodities such as silk, tobacco and sugar cane (Marks 1998, 342–3).

The 1750–1950 period saw the contraction of forests from 25 per cent to 5–10 per cent of total land area, widespread soil erosion, river sedimentation and mass declines in fish life. Ever-increasing demands on farms to produce more from ecosystems already depleted of energy reserves and with compromised nutrient cycles (due to such changes as deforestation and possibly soil organic matter depletion) led to an agricultural underproduction problem.[3]

---

3   The claim that agriculture in China suffered from acute nitrogen deficiencies (e.g. Marks

Moreover, a state apparatus debilitated by liberal democracies' imperialism and internal conflict, eventually degenerating into warlordism by the 1920s, became useless in providing even basic provisions out of collected taxes. This resulted in increasing vulnerability to famines induced by the frequent droughts associated with El Niño Southern Oscillations between 1876 and 1930, leading to tens of millions of preventable deaths. The genocidal Japanese invasion of 1937 and the horrors inflicted by warlords (variously allied to Jiǎng Jièshí's – aka Chang Kaishek's – Guómíndǎng and/or the Japanese) contributed to lasting environmental destruction that could only be worsened by the 1945–9 civil war, precipitated by an uncompromising, power-hungry Guómíndǎng supported militarily by the US. A major, rarely recalled catastrophe involved the Guómíndǎng's wartime actions. In 1938 they blew up the Yellow River's dykes in a vain attempt to stop the Japanese army's advance. The ensuing flood drowned about a million people, led to four million total deaths, displaced millions of people, set in motion wide fluctuations in the river's course, destroyed hundreds of thousands of villages and millions of hectares of cropland and, adding disaster to calamity, failed to stop the Japanese army. In other words, by the time they took state power the CPC inherited highly degraded ecosystems, a country devastated by decades of war, a farming sector in tatters, foreign market dependence even for basic goods, a militarily ravaged low-grade industrial base and a financially leaky and weakened state apparatus incapable of meeting people's basic needs (Marks 2017, 287–8).

2017, 287) must be taken with much care. What is regarded as enough soil nitrogen levels varies according to cropping system and production aims. If the demand is to produce large quantities of nitrogen-demanding crops for commerce, then it is likely that soil nitrogen levels will quickly become insufficient. Such a situation can turn catastrophic when farmers are displaced or killed through wars, overwhelming taxation and droughts, as was often the case in China especially after the establishment of the republic in 1911 and the demise of the Qing Dynasty. However, soil nitrogen levels can be manageable in the absence of such terrible situations if crops are less nutrient-demanding and production is mainly for subsistence, and if soil organic matter, where soil nutrients are mainly held, is replenished with the likes of green manure and other organic techniques not requiring agrochemical inputs. It is therefore more likely that the soil nutrient problem was due to declines in soil organic matter, rather than only soil nitrogen. There are anyway insufficient data from the period to make estimates on soil nitrogen levels.

Environmental Impacts and Policies under the Maoist Fraction of the CPC

Shortly after the revolution the US imposed a trade embargo and sanctions that impeded any import of technologies or other products, including agrochemical fertilisers, which could have helped alleviate the suffering of millions in China after such trying decades. The CPC are hardly to be exempted from criticism though. Relations with the USSR could be justifiably tense, and they certainly were by the late 1920s (see Chapter 2). The split with the USSR over political priorities (the worst of it happening over 1956–66), including the question of international leadership relative to mainstream communist parties, was entirely preventable, as was the 1969 border conflict. The split with the USSR only made matters worse overall, including by way of undermining prospects for technological assistance, especially in farming. Under such circumstances, imposed and self-inflicted, it would have been a remarkable act of self immolation for any political formation in power to leave matters as they stood, instead of doing the utmost to establish the basis for producing consistent, reliable surpluses to house, clothe and feed people.

Agricultural expansionism and accelerated industrialisation were not too different strategically from that employed decades earlier in the USSR. The policies led to similar atrocities and disasters, including a major famine. But they also led to what Kenneth Pomeranz has identified as a transformation of 'the notorious "land of famine" of the 1850–1950 period into a crucial grain-surplus area' by means of a massive expansion of irrigation, which 'contributed mightily to improving per capita food supplies for a national population that has more than doubled since 1949' (Pomeranz 2009b, 9). The overarching aims were relatively straightforward: raise productivity levels (that is, industrialise) as fast as possible to hasten the establishment of the material conditions necessary for socialism (in line with predominant understandings that egalitarianism is not feasible on an empty stomach) while parrying threats from liberal democracies and then also the USSR (Shapiro 2001, 2).

However, Máo Zédōng and associates did not have the level of social influence Stalin had enjoyed (and usurped). Máo's fate, increasingly having to struggle against being marginalised by capitalist roaders (a main motivation behind the 1966–76 Cultural Revolution), was more akin to that of Khrushchev. His failures over attempts to compete with the US over, among other pursuits, agricultural productivity cost him his post, even as it was under Khrushchev's government that the marks of full industrialisation were lastingly chiselled into many landscapes and in the minds of most people in the USSR.

Industrialisation in the PRC, in contrast, remained hampered by a dearth of locally available resources. Irrespective of such a disadvantage, machinery, agrochemicals and hybridised crops, among other means of production, were developed and output did increase (Eisenman 2018; Xu 2018). Problems of environmental destruction were noted with much concern, but were not given the same weight as economic output. Still, between 1956 and 1957 sanitation standards were promulgated for all industry and water systems, along with water quality protection and soil conservation policies. These would guide much of the environmental regulation until the 1970s (Muldavin 2000, 252).

Resorting to expanding farming to even more land, leading to more uncultivated ecosystem destruction, and pro-natalist policies to increase the numbers of workers may seem perverse in hindsight. The reality of a decimated population and economic isolation should temper such judgement. Be that as it may, the imposition of large, centralised farm units on peasants ('collectivisation') compensated for low productivity levels and amassed the necessary capital to industrialise rapidly (Eisenman 2018). A result was agricultural encroachment into much grassland in Inner Mongolia, Jiansu and Xinjiang, and deforestation in some mountainous and tropical areas, especially in Manchuria and Yúnnán. Within 30 years (1950–80) arable land increased from 80 to 130 million hectares (Marks 2017, 310–12, 320).

This may at first seem very destructive until it is realised that by the middle of the 1970s reforestation efforts were successful enough to increase wooded area to 12 per cent of land cover, winning the praise of at least some world-renowned conservationists (Eckholm 1976). Tens of millions of hectares of land were replanted with trees through massive socialist state reforestation campaigns started in the 1950s, which continue today. To a minor extent forest recovery also occurred by setting land aside for spontaneous regeneration. Grasslands and shrubland, however, have contracted overall, mainly due to cropland expansion. Claims of inflationary official figures over the extent of reforestation and afforestation (e.g. Marks 2017, 320–1) have been disproven by a recent NASA-sponsored study that combines multiple sources of evidence (Liu and Tian 2010). At the same time, inadequate attention to local ecological contexts has resulted in afforestation worsening conditions relative to biodiversity and water supplies in some areas (Shixiong et al. 2011). Still, the reinstatement and expansion of forests is a major feat of generally constructive biophysical impact while at the same time expanding food production potential.

According to an analysis of soil quality trends from the 1930s through the 1980s, no net soil erosion occurred countrywide, but there were major regional topsoil loss problems of lasting consequence (in the loess areas, for example). Though nitrogen and soil organic matter did decline (as they have in virtually all major industrialised countries), potassium and phosphate levels increased (a typical result of decades of chemical fertiliser input) and alkalinity and salinisation problems have been kept largely in check. Urban expansion, in contrast, is wrecking much nutrient-rich arable soil, even if to a geographically very confined extent (Lindert 2000).

Shifting to the State Management of a Capitalist Economy in the PRC

The environmentally worst was to come with the completion of industrialisation after the capitalist-friendly policy changes instituted in 1978, after the ousting of Máo's CPC faction. This was by no means a foreclosed outcome. Environmental protection efforts, gathering momentum under Máo's leadership with the reforestation and afforestation campaigns, were enshrined into law in 1978. The PRC's National Environmental Protection Agency was instituted shortly thereafter (Wu 1987). But the farming improvements, the industrial base and moderate environmental successes built through much sacrifice were redirected towards profit-making ventures and intense capital accumulation for the few. Centralised control over environmental practices was substantially reduced. Land and factories were mostly privatised, state planning and control were drastically reduced just as new environmental regulations were to be enforced, zones of even more intense worker exploitation were set up (special economic zones, such as Shēnzhèn, near Hong Kong) and the national economy was increasingly opened to investment and imports from the capitalist countries. Environmental regulators remained largely uncoordinated and environmental monitoring and assessment remained embryonic.

Between the 1980s and 1990s the real great leap forward occurred. Levels of industrialisation comparable to those of the US were achieved in the PRC within just two decades. This was not out of the blue, though. Agrochemical plants and production had expanded markedly beginning with Máo's waning days of leadership, thanks to friendlier terms established with the US Nixon administration. Not long afterwards, agricultural output soared along with agrochemicals exports along with farmland and water quality degradation. Land use intensification behind higher farm output at the village 'commune' level (nominally collective farm units remunerated increasingly through wages instead of work points) was achieved by increasing surplus extraction from peasants and at the cost of rising pollution problems (Eisenman 2018; Muldavin 2000).

There were and have been many and major negative environmental consequences to these shifts, including wetland and grassland contraction and an unprecedented increase in the emission of greenhouse gases surpassing those of the US, the now former number one polluter of the atmosphere (Liu and Diamond 2005; Smil 2015; Zhang et al. 2015). After a steady rise mainly since the 1970s, barring the blip caused by the Great Leap Forward industrialisation drive (1958–62), CO_2 emissions started spiking exponentially in 2002 and in short order arrived at the world's leading position, as examined later in this chapter. Controlling for population size, though, the figures are still less than half those of the US.

Air pollution has been horrific in major cities, with sometimes a series of days of impenetrable haze with such strong specular reflection as to give the impression of living within a large, shiny, suffocating envelope. Much of the problem was due to coal use in urban centres, which has been greatly worsened by emissions from the rising fleets of motorised vehicles. Air quality resembles what had happened in the wealthiest capitalist countries by the 1960s, with a high average degree of fine dust (89 μg m^{-3}), sulphur dioxide (48 μg m^{-3}) and nitrogen dioxide (34 μg m^{-3}) pervading the air of 113 cities. In rural areas, burning fuelwood and coal is also a major health hazard (Kan 2009).

Over the past decade, however, some water quality improvements have occurred in rivers and lakes, with biological oxygen demand falling by about 3 per cent. This has been part of a ramping up of investments in environmental protection, which have trebled from 0.5 to 1.5 per cent of GDP (Kan 2009). Smog has been successfully and markedly reduced, mainly by investing more in end-of-pipe technologies, including desulphurisation, increasing reliance on methane and by much more effective enforcement of regulations over industrial emissions and vehicular traffic (Mao et al. 2014; van der A et al. 2017; Ying and Carlowicz 2016).

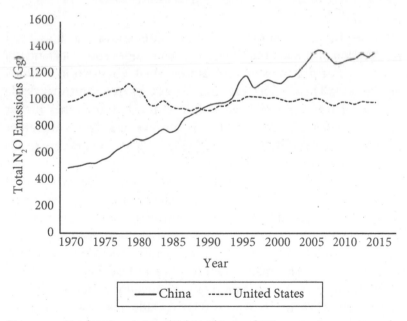

Figure 4.7   Total N$_2$O emissions (Gg) in China and US, 1970–2014
Source: EC-JRC (2020).

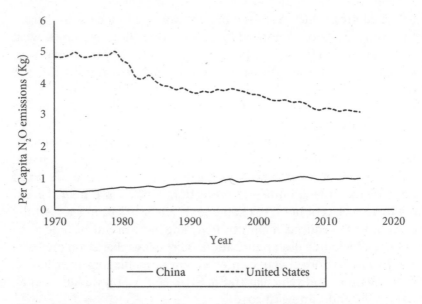

Figure 4.8   Per capita N_2O emissions (kg) in China and US, 1970–2014
Source: EC-JRC (2020).

But this has been mainly confined to cities and was a short-lived amelioration at the national scale. Nitrous oxides rose substantially in the 1990s, surpassing the US, among the highest world emitters (Figure 4.7). Then, after reaching a peak by 2007, nitrous oxide emissions have been stabilised at a level nearly three times higher than 1970. Taken on a per capita basis, however, the figures are rather low, especially compared to the US (Figure 4.8). In part this rise is due to more motorised transport use.

Sulphur emissions (Figure 4.9) follow a similar pattern, except that in the US they have steadily declined since 1979, reaching levels not experienced in China since the early 1970s. This remarkable feat is the result of the above-discussed successful 1979 UN long-range pollution treaty, which was realised thanks to the USSR. Total fine particulate matter emitted has instead been very high, mainly through coal combustion later primed by vehicular exhaust, aside from the expansion of industries, including via relocation from abroad (Figure 4.10). As part of the result of switching from coal, total methane emissions have tended to be high (higher than in the

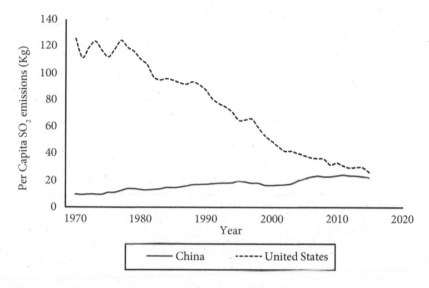

Figure 4.9   Per capita SO₂ emissions (kg) in China and US, 1970–2014
Source: EC-JRC (2020).

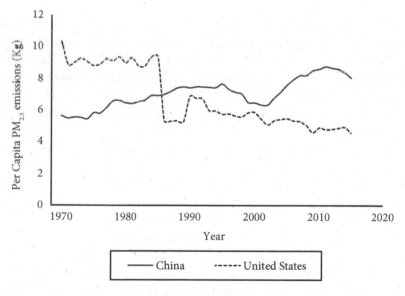

Figure 4.10   Per capita PM₂.₅ emissions (kg) in China and US, 1970–2014
Source: EC-JRC (2020).

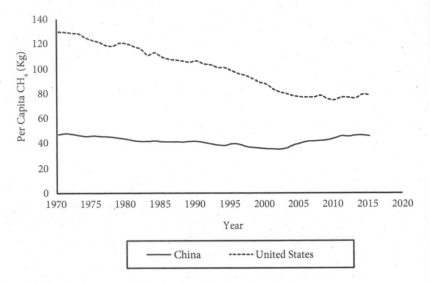

Figure 4.11   Per capita CH$_4$ emissions (kg) in China and US, 1970–2014
Sources: Crippa et al. (2019) and EC-JRC (2020).

US) and have further increased since the early 2000s, but they are still much lower on a per capita basis (Figure 4.11). On per capita terms much of the air pollution remains lower than that of the US, save for PM$_{2.5}$ emissions, an indicator of the extent of at least potential smog conditions. Sulphur emissions are also approaching the levels of the US, as a result of increasingly higher rates in China and declining ones in the US.

Soil pollution has intensified and expanded to such an extent that by 2014 above-standard trace element concentrations affected 16.1 per cent of the total land area (Chen et al. 2016). In many of the main cities, aside from smog problems, the levels of water contamination are high or risky enough that people must procure or purchase bottled water to drink and wash food (as the author also experienced). If it is any consolation, at least all this is well known because of effective environmental monitoring, and measures are being taken to contain the problem, including harsher and clearer legal frameworks, committing to ten-year national soil surveys and setting up monitoring stations making available continuously updated results (Li et al. 2019;

Zhou and Liu 2018). This is hardly enough, especially in terms of prevention, but China is on its way to a level of soil monitoring that used to exist in Hungary until 1989 or that exists to a similar extent in the Netherlands. In the US, instead, as the US Environmental Protection Agency states, 'Currently, no single information source tracks the extent of contaminated land nationwide' (US EPA 2020b). In China, at least, they know where they can start tackling the problem.

Meanwhile, 40 per cent of mammalian species and possibly 80 per cent of plant species are threatened with extinction, while hundreds had already been lost since before the 1949 revolution (Marks 2017, 330–6). These are but some of the negative impacts that have occurred or have been intensified since 1978, including major river water diversions, large-scale damming (the Three Gorges Dam being a prominent example), air and soil pollution linked to mining, industrial plants and the dizzyingly fast diffusion of motorised vehicles (Chen et al. 2011).

Not all impacts have been as terrible as usually depicted (but this can be said of the wealthiest capitalist countries too). Deforestation, while it ran amok again in the 1980s, was somewhat reversed by the 1990s. The increasingly privatised 'collectivised' forests were felled liberally, overwhelming the timber market in such a way as to reduce logging, ironically, in state-owned forests (amounting to about 42 per cent of total forest cover), where historically much damage had been done to provide fledgling manufacturing and building sectors with raw materials before the late 1970s. Regrettably, in the 1990s, with even more liberalisation of the state forestry sector in supplying businesses, the rate of deforestation, including old-growth survivals, increased in state-owned forests as well. Such was the increasing magnitude and frequency of flooding, landslides and other disastrous effects, causing thousands of deaths, that a logging ban was put into effect in 1998.

In some parts of China reforestation efforts have since been successful, such as in Yúnnán, but in part the logging problem has been displaced to other countries, as China has become an importer of lumber, resulting especially in tropical forest losses elsewhere. This possibility of halting environmentally harmful practices, at least, can be counted as a merit for a one-party socialist government that is supposed to promote socialism while ruling over a capitalist economy (which is, in a way, as contradictory as social democracy). Biodiversity prospects have also been improving with the establishment of 2,000 nature preserves by 2003, totalling 14.4 per cent of land area (Liu and Diamond 2005, 1183).

But the net effects of CPC and business decisions have been destructive, which points to major failures on the part of the capitalist-oriented ruling fraction of the CPC to restrain or suppress environmentally destructive businesses. Desert expansion has been increasingly problematic, even if it dates as a problem to periods preceding 1949. The problem in figures is the increase of areas classified as desert from 15 to 25 per cent of total land area, a startling outcome driven especially by the loss of grasslands to industrialised farming, mining and other impacts of an extractive character. The problem has been exacerbated by groundwater withdrawals at faster than regeneration rates.

An Overview of Environmental Impacts in the PRC

The socialist state phase of the PRC was short, lasting between 1949 and 1978, barely three decades. Generally, from the 1950s through the late 1970s, the prevailing Máo fraction of the CPC introduced sanitation standards, included environmental concerns in town planning and expanded forest cover. They oversaw a large-scale reforestation campaign that helped re-establish much of the forest cover lost over prior centuries. Environmental legislation became part of the constitution by 1978 with the 3rd Plenary Session of the 11th Central Committee of the CPC. This laid the groundwork for the creation of the National Environmental Protection Agency in 1988.

Nonetheless, the capitalist-oriented reforms under Dèng Xiǎopíng led to soaring greenhouse gas emissions, some desert enlargement (also in part due to climate change), an increase in species under threat of extinction or going extinct, major expansion in urban areas at the expense of surrounding ecosystems and the spread and intensification of soil, air and water pollution. After a brief spell of deforestation, however, campaigns to regain forest cover resumed and have been largely successful, alongside air pollution control measures over the past decade. Investments in renewable energy have also transformed China into a vanguard of more environmentally friendly alternative technologies. Arran Gare (2012, 14–15) captures the essence of the systemic change and its repercussions most effectively:

> China is best understood as a capitalist economy going through the stage of primitive accumulation with the state serving the interests of the bourgeoisie. In this regard, its pre-eminent concern, essentially, is to assure high profitability of business enterprises by dispossessing the peasantry of their means of production in order to create a pliable proletariat in the cities. It is more complicated than this, however. Transnational corporations have been the greatest beneficiaries of the cheap labor and lax environmental regulation. The local bourgeoisie use everything at their disposal to increase their share of profits but often work on very low margins and are very susceptible to changes in the global economy. The central government also makes some effort to ameliorate the conditions of employees. However, the involvement of local government and the fact that local government performance is defined mainly in terms of GDP growth and therefore primarily oriented to attracting and maintaining foreign investment, along with a corrupt and ineffective legal system, means that the government effectively sides with capitalists against employees and local populations. To compound the situation, it is becoming evident that many officials and Communist Party members are leading capitalists and are aligned with international capital rather than local capital.

So, overall, the shift to a capitalist economy has been an environmentally disastrous turn, even if most people in China, thanks to state interventions and limited social welfare policies, have generally benefited with higher living standards. Those benefits must not be underestimated but they are undermined by extreme economic inequality and the erosion of biophysical conditions. The total EFC steadily increased since the 1960s and then took a steep rise in 2002, surpassing by 2004 the country that until then had been the biggest drain on the planet, the US. However, on a per capita basis the US has at least twice the impact of China (Figure 4.12).

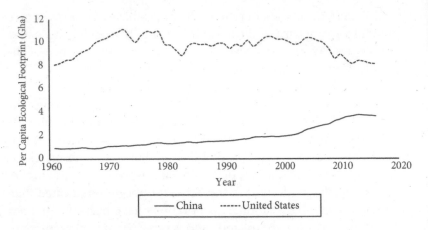

Figure 4.12   Per capita EFC (Gha) in China and US, 1961–2016
Source: Global Footprint Network (2019).

Acutely aware of this contradiction between raising well-being while destroying its social and ecological basis (see also O'Connor 1988), the PRC government has aimed, officially since 2007, to build an 'ecological civilisation', a phrase coined by a Chinese agricultural economist in a 1984 Russian publication and etymologically tied to 1970s intellectual discussions within the USSR. Promisingly, this goal of moving beyond industrial civilisation was written into the country's constitution in 2018. It remains to be seen whether the push for environmental sustainability is coupled with socialism and whether policies and new environmental practices overcome the amount of damage already done and still being done. At the same time there is now at least institutional legitimacy for ecologically minded communists in the PRC struggling to overturn the capitalist system (Gare 2020; Huan 2016).

The PRC in International Context

It is relatively easy to lay the blame on the CPC (or the nominal 'socialism' they stand for) for globally deleterious effects of industrial output and increasing consumption rates in the PRC. After all, they rule the country. Leaving the matter there overlooks the fact that the

CPC hardly have control of the economy as a whole, and this aspect is important if one wants to find out where to bring pressures to change course. Just as noteworthy is the decentralised nature of economic activity and regulatory environmental enforcement since the 1990s, where capitalist pressures and local officials' economic incentives degenerate into a lack of environmental policy enforcement (Tao et al. 2019). Most businesses are in private hands, including foreign ones. Many enterprises, whether state owned or privately owned, can and do go bankrupt, as in any other capitalist system (Huang 2012). If all firms were controlled by the CPC one would expect especially state-owned enterprises to be immune from liquidation. They are not and workers get sacked, in droves. Substantive change towards ecological sustainability and real socialism would then have to include targeting all the private domestic and foreign firms benefiting from the above-described ills. Nevertheless, the ruling capitalist faction of the CPC is certainly blameworthy relative to policies instituted and carried out.

More importantly, there is also such a thing as class struggle in China. There have been thousands of strikes and small revolts for several decades over working conditions and also over environmental quality (Li 2017), as well as peasant resistance to village communal land privatisation and rural protests over environmental degradation (Muldavin 2000; Jing 2010). Furthermore, there persist divisions within the party and among party-aligned capitalists. Within the CPC there do remain genuine communists, many of whom are environmentalists, such as Pan Yue. These fractions of the party have been steadily sidelined, however, after some promising movements in the early 2000s (Gare 2012).

There are in any case international linkages to be considered. Were the CPC capable of controlling all industries, which they of course do not, one would still have to account for the massive foreign direct investment propelling industrial production since the 1980s and the presence of industrial plants operated by private firms from countries like the US or Germany. In a highly intensified global commercial interlinkage of countries, especially since the late 1980s, one cannot pretend that such human-derived calamities as pollution and species extinctions within and from the PRC are the product of the PRC alone (this goes for almost any country, even the US). As pointed out above, comparisons between countries are a dubious undertaking when countries' contexts are not considered. It is, for example, because of the increasingly dense commercial interlinkage across continents since the 1980s that the PRC has become a net exporter of energy, even if a net oil importer since 1993. This paradox is due to the energy embodied in the manufactured commodities exported (Chen et al. 2011; Hongtao et al. 2010; Huang et al. 2019). At the same time, if one accounts for total ecological footprint (equivalent yearly biological production – measured as global hectares – needed to sustain given resource consumption rates), the PRC has been a net importer of energy and in net ecological deficit since at least the mid-1990s, barring the 1997–9 period (Li et al. 2007). The PRC is as environmentally unsustainable as basically all the liberal democracies, as already shown in the comparative analyses in Chapter 3.

What this relative unsustainability hides, however, are the global capitalist interconnections that enable it. That is to say, the ecological devastation in China is at least in part predicated on linkages with the most powerful capitalist institutions and states. The arrangement is well known to world systems analysts, who have shown empirically that semi-peripheral (middle-income) countries in the capitalist world economy tend to become sites of higher environmental damage intensity (Roberts et al. 2003) while in part offloading, through raw material imports, other forms of environmental destruction to the global periphery, or lower-income countries:

China, a middle-income nation, is exploiting forests of other low- and middle-income nations in a manner similar to high-income nations as it tries to protect its own forests while meeting its needs for wood, fuel, paper, and pulp from abroad to stimulate its economic growth and satisfy its changing consumption patterns. (Shandra et al. 2019, 100)

This can explain why reforestation has succeeded within the PRC just as timber imports have skyrocketed since roughly the early 2000s. Otherwise stated, if the world's 'chimney' (a rather terrible metonym for the PRC; as in Malm 2012) is to provide endless stuff for endless profits there must be endless worker exploitation (wages) and endless amounts of fuelwood costing much less than the stuff the chimney produces for the world market. That lower-priced fuelwood is predicated on chains of exploitation of all sorts of life forms in a highly racialised global pyramidal structure of power.

This semi-peripheral chimney situation for the PRC has now been in place for decades, arguably emerging out of the July 1971 US National Security Advisor's secret Beijing visit. The rest of the world, especially the wealthiest countries, get low-price manufactures, as long as Chinese workers' wages are comparatively low, while many Chinese (as so many elsewhere) suffer the consequences of pollution, involving hundreds of firms in many countries, including liberal democracies and the PRC, raking in huge profits. As Andreas Malm (2012) has explained, capitalists flocking to countries with the lowest production costs, especially wages, will stimulate spikes in fossil fuel use and thereby greenhouse gas emissions. And it is not only people who get harmed in the process. The grotesque and biodiversity-menacing trade in wildlife from the PRC and South East Asia, for which such countries have become infamous, are linked to lacklustre environmental protection and enforcement in the PRC and demand pressures mainly from the European Union and Japan (Nijman 2010) capitalising, among other factors, on the low-wage labour employed to trap wildlife. Even the scientific mainstream, while not grasping the relations of production involved, recognise that 'globalisation' (the intensification of global trade and outsourcing of manufacturing) has resulted in the mass relocation of pollution from the 'developed' to the 'developing' countries, with the US–PRC relationship as the most impactful instance (Wiedman and Lezen 2018). This link is also evinced by way of unfettered global trade, which if markedly reduced would substantively decrease both CO_2 and particulate emissions (Lin et al. 2019).

One way of representing the repercussions of this arrangement is by estimating the premature deaths linked to industrial emissions of $PM_{2.5}$ (dust smaller than 2.5 microns) and their long-range transport. Perhaps the reader will recall the great attention given in the press in the early 2000s to the eye-watering giant brown cloud periodically hovering over the lands of South and East Asia. Much was made of its menace to health and amplification of global warming trends (Ramanathan et al. 2005), but it was also another chance, mostly missed, to get over any persisting institutional dichotomy between health and environment and barmy assumptions of physical environments respecting national borders (Mitman et al. 2004). Missed almost entirely in a haze of hyperbole, metaphor and chatter was how intertwined cross-regional commodity and pollutant flows are.

For air pollution from global commerce is deadly under currently predominant social arrangements and their associated environmental practices. Pollutants within the PRC are in part due to the outsourcing of industries from the US. But there is a differential boomerang effect involved as well. While the US north-east enjoys a reduction in air pollution, air quality in the western US is impacted by pollutants coming from outsourced US enterprises manufacturing their products in China (Lin et al. 2014). The broader results of capitalist globalisation, however, are truly grim. For 2007, slightly more than a third of the world's 3.45 million premature deaths from $PM_{2.5}$ pollution was traceable to production for international trade. This is greater than mortality rates linked to long-distance air pollution. Some 12 per cent of world premature deaths were associated with pollutants coming from other regions, while 22 per cent resulted from production of commodities for export. For the PRC, the picture is similarly disproportionate. Airborne pollutants travelling from export-oriented industries in China led to 64,800 early deaths elsewhere, with 3,100 in western Europe and the US. In exchange, 108,000 Chinese died prematurely from the air pollution emitted to produce commodities bought in western Europe and the US (Zhang et al. 2017). Not only that, until 2018, when a ban was legislated, China was the destination for 70 per cent of the world's electronic and plastic waste from the largest economies, like the US (Frey 2012; Kan 2009). The waste-producing consumption of goods abroad is linked to the distressing health effects of domestic polluting industries and international dumping. In some ways it is a situation like Cancer Alley in Louisiana (US), which is in some measure a product of US petrochemical exports that enable firms elsewhere in the world to manufacture other commodities for export.

Another major illustration of this uneven and combined trade in devastation is related directly to the now decades-long planetary climate catastrophe. Global trade has risen from 27 to 60 per cent of world GDP between 1970 and 2019, with the PRC playing a pivotal role (World Bank 2020a). By 2004, international trade took up about a quarter of global CO_2 emissions, mainly through exports from the PRC (Davis and Caldeira 2010). Trade between the US and PRC alone contributed about 7.2 gigatonnes (Gt) of CO_2 to the atmosphere between 1997 and 2003. According to US Department of Energy data this was a small fraction (0.004 per cent) of the 173.9 Gt world total, but still close to twice Vietnam's 3.9 Gt, for example.[4]

Much is being made of the PRC surpassing the US in 2006 as the foremost CO_2 emitter, but what is astonishing is how much more impacting greenhouse gases like N_2O and hydrofluorocarbons (HFCs) rarely form part of such a discussion. Yet N_2O (nitrous oxide, laughing gas) has nearly 300 times as much atmospheric warming potential as CO_2 and HFCs hundreds to thousands of times more (UN Climate Change 2020). Not only that, N_2O is a stratospheric ozone-depleting substance not covered under the Montréal Protocol (Ravishankara et al. 2009), a treaty already hobbled by methyl bromide exemptions to ingratiate large corporate farming interests (Gareau 2013). Add the glaring laughing gas omission and the much-inflated poster child of international environmental agreements treaty groans under the weight of raw capitalist deception. It turns out that total N_2O emissions in the PRC overtook those in the US by 1990, though not in per capita terms (Figure 4.7), and without the fanfare allotted to the same outcome observed with respect to CO_2. The US, however, remains in first place relative to total HFC emissions with about 301 million tonnes of emissions, at least as of 2010. The PRC is a distant second with 184 million tonnes, after overtaking Japan's place by 2000.[5] None of this makes any of these countries' ruling classes any friendlier to the ozone layer. There seems instead to be widespread indifference among them about millions of people getting ever more quickly seared by carcinogenic ultraviolet radiation. This is an indict-

---

4   https://cdiac.ess-dive.lbl.gov/.
5   World Bank, https://data.worldbank.org/indicator/EN.ATM.HFCG.KT.CE, accessed 22 October 2020.

ment of capitalist states as well as one-party socialist governments presiding over capitalist economies, such as the PRC.

N_2O emissions in lower-income countries like China are tied to levels of foreign direct investments, as shown for a study covering the years between 1990 and 2014 (Mejia 2020). Reducing such economic dependency could improve matters. This could be similarly argued about CO_2 emissions, about which there has been much research. Over the above-discussed 1997–2003 period, US CO_2 emissions would have been 3–6 per cent higher without importing goods from the PRC, while exports from the PRC to the US accounted for 7–14 per cent of the PRC's CO_2 emissions (Bin and Harriss 2006). In 2003 there were 4.5 and 5.9 Gt of CO_2 emitted from the PRC and US, respectively. Adjusting for trade effects using even the lowest percentages (that is, in the US's favour), the US would have emitted 6.2 Gt and the PRC 4.1 Gt of CO_2. But by 2018 the PRC's contributions far surpassed those of the US. The figures are about 10.1 Gt for the PRC and 5.4 Gt for the US. At the same time, trade between the two countries has increased fivefold in monetary terms since the above-cited studies, according to the US Census Bureau.[6] The difference has remained about five to one in favour of exports from the PRC, but the volume of commodities traded has markedly increased. The oft-cited figures therefore exaggerate the divergence in annual emissions between the two countries. Using the higher percentage from Bin and Harriss (2006) to adjust for trade effects (14 per cent for 2003), the resulting 2018 totals make the difference of 5.7 and 8.7 Gt CO_2 emissions between the US and PRC much less dramatic than what is often quoted. When accounting for net international trade effects (Ritchie and Roser 2017) a similar outcome emerges. For the US, a net importer (i.e. net consumer of world resources), about 8 per cent should be added to the 5.4 Gt figure, yielding 5.8 Gt. Conversely, about 13 per cent should be deducted from emissions in the PRC, a net exporter, resulting in 8.8 Gt.

More broadly, in a wider global context commodity production in China responds inordinately to foreign demand relative to the US. Roughly a third of the PRC's GDP is accounted for by exports

---

6   www.census.gov/foreign-trade/balance/c5700.html, accessed 21 October 2020.

(at least prior to the Covid-19 pandemic). In 2005, CO_2 emissions due to exports exceeded import-related emissions by 1,000 million tonnes (Lin and Sun 2010). Since joining the World Trade Organisation in 2001, foreign direct investment and commerce have so intensified as to usher in another major upwards shift in CO_2 emissions (Ren et al. 2014), the other having been reached thanks to the changes brought about in 1978. These shifts are evident from the two instances of steepening slope inclinations in the PRC's cumulative CO_2 emissions curve, one after 1978 and the other after 2001 (Figure 4.13). Importing goods produced by outsourced industries makes the effectively highest resource-consuming countries in the world, like the US or UK, look environmentally better or even like they are reducing emissions. But if CO_2 embedded in imports were included in greenhouse gas accounting, UK figures would be twice those typically reported (Harvey 2020).

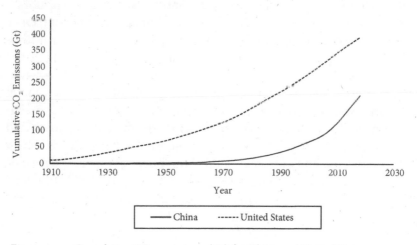

Figure 4.13   Cumulative CO_2 emissions (Gt) for China and United States, 1910–2018
Source: Ritchie and Roser (2017).

Still, export-heavy production and foreign investments are only two factors behind China's greenhouse gas emissions record having attained such dismal heights. Emissions keep rising at alarming rates for many other reasons, including profit-oriented economic policies, sustained coal dependence and technological challenges. However, halting a critical source of national revenue – to expand the internal market, introduce or renovate basic infrastructure, develop more environmentally friendly technologies and lift up people's standard of living, among other objectives – is not much of an option unless world powers, like the US and its NATO satellites, collaborate with the PRC in achieving a global green transition (Lin et al. 2019). To understate the issue, this is unlikely in a capitalist framework, so the PRC will continue to seem like the main culprit and increasingly like a necessary priority target of green reform.

That is, until one acknowledges still other ways of looking at the same data aside from total yearly emissions. When population size is included in the analysis, the PRC emits less than half (7 Gt) of the CO_2 emitted in the US (16.6 Gt). This may make little difference to atmospheric dynamics, but even when just focusing on the atmosphere countries like the US are hardly exonerated. Nor does it make the PRC necessarily steal the carbon emissions limelight. Once CO_2 is airborne it takes decades to a couple of centuries for it to be scavenged from the atmosphere and stored away in some other part of the planet (or geosphere). The same story goes for other greenhouse gases, save for methane (about a decade of residence time in the atmosphere). Looked this way, the location of the highest greenhouse gas emissions changes again. Historically, more than 405 Gt of CO_2 have been emitted in the US, attaining the world record by far, even if the emissions curve displays much less spectacular slope changes in the early 1970s and early 1990s (Figure 4.13). In China (including now the PRC), in 2018 a distant third in the ranking, the cumulative total is 210 Gt, with the then EU-28 not far behind the US in second position with 356 Gt.

In contrast to the US and other countries that developed into capitalist formations, colonised and plundered the rest of the world, and thereby got industrialised early, the PRC shook off colonising powers late and started off in a situation of mass poverty that was overcome by industrialising rapidly, as in the case of the USSR. The CPC Dèng faction's 1978 takeover and the subsequent conversion of the PRC into a capitalist machine marked the start of a massive expansion of industrial capacity. Given its basis in fossil fuels, the industrial expansion led to an exponential rise in greenhouse gas emissions. The major shift in the PRC occurred as global warming, tropical forest destruction, biodiversity declines, ozone layer disruption and neocolonial reconfigurations (neoliberalism) were already in full swing, alongside the development of consumption technologies of unprecedented destructive efficiency and policies characterised by, among other trends, a complete abandonment of the precautionary principle (Gareau 2013), industrial outsourcing, general suppression of workers' wages, mass privatisation of public assets, bloated capitalist coffers and militaries, and ever extreme financial speculation and volatility. In this overall systemic shift towards a capitalist economy the PRC was affected by and eventually effected similar planetary environmental damage by largely following prevailing economic policies, while adding some protective buffers from speculation and from foreign corporate encroachment (as the main imperialist powers have done historically).

But for a country with, at first, the largest population and without colonies to ransack, it should not be surprising that the drive to complete industrialisation and compete successfully (and defend oneself) against world powers, while raising living standards, would result in environmental catastrophes mainly within the country and in skyrocketing greenhouse gas emissions. Not surprising does not mean legitimate, as there are always alternatives worth exploring. Any promising changes towards diffuse worker and peasant control of the economy and towards ecological sustainability were stifled first under the Máo administration and then under Dèng and succeeding governments (Gare 2012). The shift to a capitalist economy in the PRC has exacerbated matters monumentally and with great speed. It demonstrates to the rest of the world, especially the now tiny state-socialist world, the grave environmental consequences of a move towards capitalism.

However, trying out alternatives to capitalism or state socialism carry extreme risks in a capitalist world-system, such as succumbing to invasions or a coup orchestrated from abroad. Or turning into an oligarchs' playground or into an even more centralised form of despotism with a capitalist economy, as in what emerged out of the USSR bloc after attempts to build social democracy in the 1980s. Or, worse, a state-socialist country could descend into the horrors of civil war, get torn up like Yugoslavia, and then become subordinated to the dictates of a foreign capitalist power. These outcomes have been neither socially nor ecologically positive. They did not go unnoticed by the CPC leadership and convinced those who had prevailed by 1976 within the CPC to reinforce their political grip in overseeing the capitalist road already taken by 1978 (Lane 2014).

When it comes to global impacts like climate change it is also worth considering the issue of the long-term scale of impact per country. Addressing greenhouse gas emissions while raising living standards for millions, as they still need to be, should entail having countries that have stuffed the atmosphere the most with greenhouse gases not only phase out fossil fuels and other greenhouse gas sources soonest, but also overhaul land use so as to have net absorption and storage (sequestration) of greenhouse gases for the next hundred years or so. The same should be argued about economic and technological overhauls to address historical and current world deforestation, biodiversity loss, soil destruction and various forms of pollution, especially in the oceans, among other environmentally destructive impacts. And this massive redirection of funds is without even demanding historical reparations, global wealth redistribution and the free sharing of environmentally constructive technologies and techniques. For the US, in terms of impacts on the atmosphere, it could be argued that twice as much of these kinds of changes need to be made compared to the PRC to undo the long-term harm done. At the same time the PRC inherited greenhouse gas emissions (estimated at 1.8 Gt of CO_2) for which other countries are also in part responsible (the UK is especially complicit alongside others partaking of the colonisation of China: Austria-Hungary, Belgium, Czarist Russia, France, Germany, Holland, Italy, Japan, Spain and the US). Although such pre-revolutionary emissions are infinitesimal compared to the massive atmospheric outlays from the PRC since 1949, the historical trauma of colonisation contributed to and still enables the ruling capitalist fraction of the CPC to justify economic growth and thereby military might for self-defence, with huge environmental consequences. Undermining that excuse takes much more and is much more complicated than sweeping the CPC from power.

Concentrating on the effects of greenhouse gas emissions on earth systems or on environmental devastation within the PRC without accounting for global interlinkages, imperialist pressures and histories of colonisation amounts to pretending that there exists an international level playing field and to denying the effects of highly uneven power relations. The implication of using only annual emissions totals or other environmental impact indicators to compare countries is especially perverse. It is to downplay, if not ignore, the vast disparities between countries and the significance of people's quality of life, which largely depend on energy derived from fossil fuels, precisely on account of the capitalist world dynamics that spawned and enforce such dependence. Planetary environmental damage is not addressed, much less resolved, by blaming those who are worse off for decisions and actions taken by the better off and most powerful. Surely the enormous gaps in wealth and political power should be central concerns in the struggle for environmentally constructive, life-affirming impacts. When it comes to planetary environmental catastrophes, those insisting on locating problems within individual countries or even single political parties ought to have a look at the much broader capitalist world picture to grasp the main, systemic causes that need to be addressed and, just as important, identify those political forces that wield the greatest power at the global scale. In the current conjuncture, barking up the China tree is misplaced, if not delusional. To borrow and somewhat diverge from Minqi Li (2016), the planetary-level impacts and the trajectory of the PRC, as the world centre of capital accumulation, hinges on the outcomes of global capitalist contradictions. A crucial determining process is what results from current class struggles within China and their interlinkages with class struggles in the rest of the world. A high degree of international commercial interdependence underlies greenhouse gas emissions and other kinds of environmental harm. This does not exonerate the ruling classes engaging in such activity, much less those capitalist fractions of the CPC governing the PRC. However, knowledge of the extent of environmental damage emerging from international capitalist commercial linkages reveals possibilities for reducing such atmospheric emissions by delinking from capitalist countries or linking with specifically socialist projects instead (including within capitalist countries).

CUBA: THE MOST ENVIRONMENTALLY SUSTAINABLE COUNTRY ON EARTH

By the 1950s, cane sugar plantations had some 400 years of history on what is now known as the island of Cuba. It was a history of socially and ecologically devastating impacts, including genocides and slavery. Less than 30 per cent of forests remained because of cane sugar production (Whittle and Rey Santos 2006). Such impacts from succeeding settler colonial regimes were behind Cuba becoming at one point the largest sugar exporter (Monzote 2008; Tucker 2000, 27).

The parlous environmental situation was magnified after US conquest in 1898, when a US client regime was imposed. Sugar cane production was industrialised and expanded into other areas of the island, bringing more deforestation, habitat losses (leading to declining biodiversity), water depletion (75,000 litres per tonne of sugar cane) and pollution, and soil erosion and related compaction, but also new problems related to agrochemicals and mechanisation, like soil acidification and heavy metals pollution (Cheesman 2004, 17–18; Scarpaci and Portela 2005, 19–20). Setting up the first national park in 1930, among very few other more positive contributions, was too little to countervail widespread environmental devastation.

What is more, sugar cane plantations were tightly linked to US firms and international sugar markets, while the island's infrastructure was geared towards raw material export. There was little to no manufacturing sector, especially not of the sort that would offer consumer items to be sold within the country. The vast majority of Cubans were undernourished, had sparse water sanitation infrastructure, little to no healthcare services, tended to be illiterate and lived in huts (O'Connor 1970; Yaffe 2019). This is the situation Cuban revolutionaries inherited in 1959, a hugely unequal, racist and largely plantation-based economy dependent on the whims of US companies, which controlled three-quarters of arable land, and their allied local upper-class despots.

Overturning Colonial and Neocolonial Legacies in Cuba

Barely a year after Cuban revolutionaries succeeded in gaining substantive independence from the US, the US government imposed a trade embargo that lasts to this day and which cut off access to the means of industrialised agricultural production and the main market for sugar sales. One outcome was the expansion of sugar cane plantations in other US colonies like Puerto Rico, as well as other Caribbean islands and Brazil and Mexico. Hence, ecological degradation was exported by the US to other countries in the American tropics (Tucker 2000, 48–50).

To make matters worse within Cuba, a drought affected the island in 1962–3. Among the initial responses was an extensive reforestation programme, the expansion in the number and extent of protected areas, and the diversification of farming with more land under cultivation, as pursued in the USSR. The government's policies of natural preserves expansion resulted in what is today a network of protected land-based, marine and coastal ecosystems. Thanks to state socialist policies, even if climate change, pollution and overfishing pose constant threats, Cuba boasts among the most well-protected marine ecosystems in the Caribbean, including coral reefs, with natural preserves covering a quarter of surrounding underwater land-masses or insular shelf (Roman 2018; Whittle and Santos 2006). In other words, from the start environmental protection and improving people's lives were not perceived as mutually antagonistic policies.

Still, to fend off starvation risks the Cuban government had to import most food, mainly from the USSR and other parts of Europe, and engage in rapid industrialisation, including farming. The resulting indebtedness forced a return to an emphasis on sugar cane export to new markets farther away to offset import-induced government deficits and to repay loans, including to the USSR and allied states. Thus, the area of sugar cane cultivation came to surpass the pre-1959 extent, affecting 30 per cent of arable land and taking up 75 per cent of export-based income. Even so, reforestation efforts were in full swing by the 1960s thanks to assistance from Czechoslovakia. Community nurseries were set up in the 1970s to conserve and share seeds and seedlings to reforest areas in the countryside degraded by plantations and mining. Over 200,000 hectares were thus reforested, contributing to a longer-term net expansion of biodiverse forest since 1959 (Rosset and Benjamin 1994; Rosset et al. 2011; Tucker 2000, 48–50).

Environmental regulation was distributed across different state offices, as in the USSR. This changed in 1976 when the State Committee on Science and Technology was tasked with coordinating environmental protection activities under the newly minted National Commission for Environmental Protection and the Rational Use of the Natural Resources (COMARNA). Such coordinating functions were folded under the Academy of Science in 1980, when the State Committee was scrapped. During the 1980s a comprehensive regulatory and action plan was developed and, in 1990, culminated in expanding the powers of COMARNA as a permanent commission independent of the ministries and entitled to enacting (contingent on state approval) and enforcing environmental regulation (Whittle and Santos 2006).

By the early 1990s, however, terrible hardships created conditions so dire as to induce the decision by the government, among other measures, to favour cooperatives and private plots and to develop and spread organic farming methods. The disappearance of the USSR resulted in the loss of crucial sources of raw materials and machinery that threatened, among other things, the ability of food provisioning and the state's capacity to ensure food access. Drastic measures (such as food rationing and conversions to low-input farming) were put in place that, after much resistance within and outside the party and state, produced much compromise among various factions in how food is to be produced and distributed, accompanied by some decentralisation in decision-making processes (Hearn and Alfonso 2012; Premat 2012). Sugar, nickel – mined through a joint venture with a Canadian firm – and tobacco still comprise the main exports, though export-oriented farming is getting more diversified. Economically, though, social inequalities are rising mainly due to the allowance of a parallel US dollar-based economy, especially in the tourism sector (Blackburn 2000, 15–18).

A decisive official turn towards ecological sustainable socialism was made evident in Fidel Castro's speech at the United Nation's 1992

Earth Summit in Rio de Janeiro, where he communicated the refusal to follow the conventional model of development, including the one promoted in the USSR. Amid a precipitous downturn that threatened to tear asunder the country, the Cuban government elected to reconfigure institutions and modify the constitution to facilitate a transformation towards an ecologically sustainable society.

In 1994 the new Ministry of Science, Technology and the Environment (CITMA) was established to succeed the Cuban Academy of Science and COMARNA. Turning a permanent commission into a cabinet-level office greatly empowered environmental regulators, who previously had less leverage relative to other ministries. CITMA was tasked with identifying inadequacies in existing environmental laws and environmental degradation problems and empowered to provide the means to remediate existing and prevent future environmental harm (Whittle and Santos 2006). The policy included promoting renewable energy, biological conservation and agroecology in food production through, among other things, educational programmes, professional training, research funding and public outreach.

In 1997 sustainable development was incorporated into the constitution (Law 81), giving environmental protection high priority (Benz 2020; Whittle and Santos 2006). In effect, existing environmental legislation from the 1970s was greatly expanded. What was novel is making sustainable development into law. It means that environmental conservation is a citizen's duty, and it is to be carried out with the well-being of current and future generations in mind. This commitment is mirrored in a proactive effort to infuse higher education with environmental literacy curricula focused on building practical skills and technical expertise (Gómez 2019). Since then even more strides have been made in developing and applying ecologically beneficial practices, especially by adopting agroecological crop production methods and developing urban farming. This has not been at the expense of other kinds of historical achievements. Advances in integrated pest management are among those feats, thanks to much state investment in research institutes. Healthcare also remains universal and research in the medical and biotechnological sciences continue to be cutting edge (Baracca and Franconi 2016; Lucantoni 2020; Navarro 1992).

Nevertheless, especially because of tremendous external and internal capitalist pressures, the road remains long in reversing centuries of environmental damage and in improving upon decades of mixed results since the 1959 revolution. Attempts to wean the economy away from fossil fuels have been made multiple times, including by means of lowering demand through low-input farming. A nuclear power option was also considered by 1967, in continuity with the Batista dictatorship's 1956 plans in concert with the US government. However, the latter period plans were with the USSR, whose downfall halted nuclear plant construction and rapidly led to its abandonment. In hindsight this was a fortuitous aspect of the massive downturn, but the project's abandonment hardly solved the energy resource problem.

Trends in fossil fuel consumption can be seen as indirectly relative to CO_2 emissions (Figures 4.14). With advantageous terms of trade granted by the USSR, CO_2 emissions climbed more or less parallel to oil imports and combustion, which also helped build up industrial infrastructure. Manufacturing and energy production at first took their toll on air quality, with rising sulphur, nitrogen oxide and fine particle emissions (Figures 4.15, 4.16 and 4.17). For the most part air quality improved with the economic downturn, but nitrogen

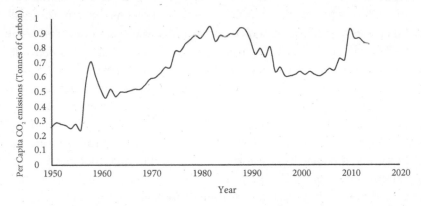

Figure 4.14   Per capita CO_2 emissions (tonnes of carbon) in Cuba, 1950–2016
Source: Boden and Andres (2014).

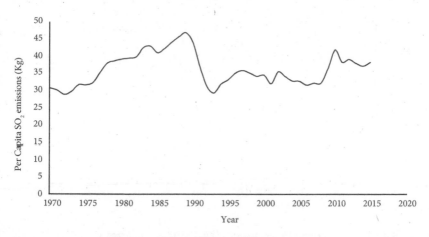

Figure 4.15 Per capita SO$_2$ emissions (kg) in Cuba, 1970–2015
Source: EC-JRC (2020).

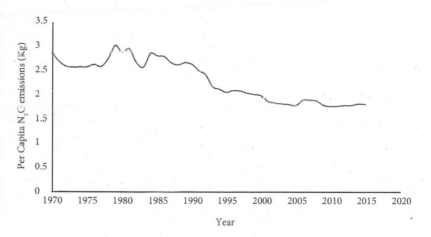

Figure 4.16 Per capita N$_2$O emissions (kg) in Cuba, 1970–2015
Source: EC-JRC (2020).

oxide, aerosol and methane emissions (Figure 4.18) had already been worked on, mainly by improving farming techniques and developing retrofitting apparatuses for manufacturing industries. Sulphur emissions persist as a problem, which necessitates more attentiveness to replacing or retrofitting aged and polluting motorised vehicles and power plants. A general problem lies in the underdevelopment of

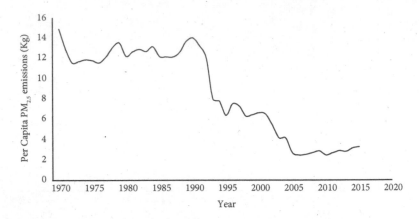

Figure 4.17   Per capita $PM_{2.5}$ emissions (kg) in Cuba, 1970–2014
Source: EC-JRC (2020).

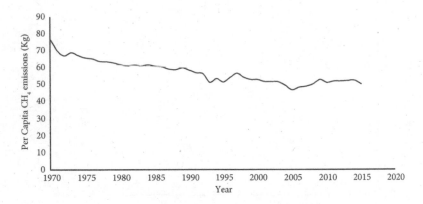

Figure 4.18   Per capita CH_4 emissions (kg) in Cuba, 1970–2014
Source: EC-JRC (2020).

railway transport or other means of transportation that would substantially reduce toxic emissions and greenhouse gases. Still, there
has been an overall positive impact on air quality that is to some
extent the result of introducing more energy-efficient appliances and
lighting and conservation measures, as well as extending the electri-
cal grid to much of the island, especially in the early 2000s.

The sharp CO_2 emissions decline in the early 1990s, due to the sudden interruption in oil flow with the USSR's disappearance, was stabilised to a lower plateau by the late 1990s. The energy procurement problem was alleviated mainly through agreements of affordable oil imports from Venezuela under the newly established socialist Bolivarian government. In the 2000s, CO_2 emissions rose again to reach the levels of the 1980s by 2014. The figures for air emissions are still very low compared to countries with living standards at or above Cuban levels. The most recent direction is anyway downward, thanks to efforts to reduce fossil fuel consumption, including by introducing more renewable energy use. But about 95 per cent of electricity is made possible by fossil fuels, so the inflection may not last long unless an infrastructural overhaul is successfully achieved (Iakovleva et al. 2020; Hornborg et al. 2019; Madrazo et al. 2018).

This is the intent of the Cuban socialist state and resources have been made available more and more towards this purpose. The 2005 Cuban Energy Revolution initiative led to the diffusion of energy-conserving appliances and fluorescent light bulbs, replacing less energy-efficient counterparts. In 2014, following through on the new energy policy, a national plan was approved to phase out fossil fuels in electricity generation, eventually replacing fossil fuels with combinations of biomass, solar panels, hydropower and wind energy sources. The long-term strategy may meet some obdurate obstacles because of the challenges of constructing the necessary infrastructure for a transition to renewable energy sources. This necessitates continuous resource extraction and labour power to establish and maintain such infrastructure, as well as accumulating enough capital and securing the resource base to do so. For island countries like Cuba, with relatively exiguous sources of energy, finding a solution is likely to entail global coordination that is attentive, relative to socialist principles, to ensuring social justice elsewhere as well as within Cuba (Baer 2018, 164; Hornborg et al. 2019). It is instead through food production, aside from biodiversity conservation efforts, that the greatest advances have been made towards ecological sustainability. Yet, overall, the far-sighted and sustained environmental measures taken up and developed over several decades have improved not only the lives of Cubans but also, as shown in the achievement of a lower ecological footprint, the state of the island's ecosystems and physical environments (Figure 4.19).

Figure 4.19   Per capita EFC (Gha) in Cuba, 1961–2016
Source: Global Footprint Network (2019).

Environmentally Sustainable Farming in Cuba

Much land in Cuba is impacted by farming (*c.*28 per cent of total land area), approximately 41 per cent of which is under cooperatives. Much land was redistributed to landless peasants and a state farming sector was instituted. State farms came to have control over 85 per cent of farmland. But in 1963 the state farm system was broken up into agricultural cooperatives, the Unidades Básicas de Producción Cooperativas (UBPC). Since 1961, nearly all smallholding farmers (those with less than 67 hectares) are organised through the National Association of Small Farmers (ANAP). Among the benefits are farm credit and interest-free loans from the state. Other smallholders have been brought into cooperatives by using incentives and persuasion, rather than coercion, as had been briefly the case in the 1970s (Rosset et al. 2011; Royce 2018).

During the 1970s and 1980s Cuba underwent the industrialisation of agriculture with the introduction of machinery, agrochemicals, hybrid crops and large-scale irrigation systems, and the expansion of monocultures. This change was facilitated by supportive measures from the USSR and their allied socialist states. The consequences were typical of this kind of high-input, capital-intensive farming, but arguably more intensely so on a high-rainfall island with a warm climate that is battered each year by hurricanes and with a long history of deforestation and widespread nutrient-depleting sugar cane plantations. Depending on landscape position, soil properties and ecosystem type these processes make for faster soil erosion, salinisation and acidification rates.

It only took a couple of decades of industrialised farming for soil erosion and compaction problems to worsen in 43 per cent and 24 per cent of farmland, respectively. Salinisation and acidification problems increased (respectively 14 per cent and 25 per cent of farmland). The combined impacts of salinisation, acidification and organic matter loss contributed to nutrient insufficiency challenges in about 45 per cent of cultivated areas. Overall, more than three-quarters of farmed area showed one or another productivity impediment (Machín Sosa et al. 2013; Rosset et al. 2011). These negative impacts, however, were already being addressed and reversed by a combination of conventional soil conservation and alternative agroecological approaches (Rosset and Benjamin 1994).

The conventional techniques have already been described relative to the USSR experience and were by and large replicated in Cuba. Novel agroecological approaches, on the other hand, have been a hallmark of the Cuban soil and agroecosystem restoration process. Much state investment in scientific education and research infrastructure facilitated this outcome, with dozens of agricultural research centres set up and dotting the country already by the 1970s. Scientific inquiry into agroecology and development of agroecological applications started within research and field centres at least a decade before the 1990s hardships. The development of biological pest management dates back even further, to the 1970s. This state-supported foray into ecologically oriented, low-input agriculture was a major factor in making possible the turn away from agrochemicals-intensive farming. It is consequently incorrect to impute the shift to ecologically sustainable agriculture to sheer economic desperation (Fernandez et al. 2018; Levins 1990).

Aside from well-known organic amendments, like livestock and green manures and municipal organic waste composting, vermicomposting has become more fully developed in Cuba. Earthworm-driven humus production has reached an industrial scale, a feat unique to Cuba. There has also been much research into and use of biofertilisers, like nitrogen-enhancing *Rhizobium* and other such bacterial inoculants, to raise soil organic matter and nutrient levels. One important innovation is the enrolment and concentration of specialised bacteria to make soil phosphorus (one of the major plant nutrients) more soluble. This way it can be more readily absorbed by plant roots. This is of inestimable value for places where soil tends to be alkaline (high pH), which causes phosphorus to be tied up by mineral particles and therefore not available to roots (Rosset and Benjamin 1994).

In 1993, under enormous economic duress, the Cuban government decided to replace many state-run farm units, managing three-quarters of cultivated land, with cooperatives run by formerly waged farmers (as state employees). Now only a third of farmland is under state farm management. As an advantage and incentive over UBPCs and private farms the newly created cooperative farms can use state-owned land in perpetuity without having to pay rent (usufruct rights) and can avail themselves of low-interest credit to purchase farming equipment and infrastructure. This is the same year when more stringent soil conservation laws were passed. The combined environmental and social policies worked. Farming impact has shifted increasingly in a net positive ecological direction while the precipitous decline in food production and availability was rapidly overcome and reversed to reach among the highest per capita food production rates in South America and the Caribbean. Undernutrition rates, which in the early 1990s afflicted a fifth of Cubans, were reduced to 5 per cent by the 2000s and to 2.5 per cent by 2014, a level comparable to high-income countries (Betancourt 2020; Rosset et al. 2011).

Other major changes resulted from socialist state efforts in conjunction with popular support. After a peak of 36 per cent of total land area in the early 2000s, farmland has contracted to the current 28 per cent figure, which is at about 1980s levels. This has allowed for expanding areas of conservation, even if not necessarily planned that way. What has been planned, however, was a continuation of the reforestation programme. This is manifested in a nearly 2 per cent annual increase in forest area since 1991 (Betancourt 2020; Rosset et al. 2011; Royce 2018; World Bank 2020b).

By the 2000s, smallholder farming in Cuba amounted to about a fifth of cultivated area. In 2015 this fraction had climbed to 40 per cent and has come to predominate tobacco and coffee production, while sugar and other export-oriented crops are mainly under cooperatives. Another major shift occurred by 2002. As part of a crop diversification effort, sugar cane production was reduced by roughly a third (mainly by closing sugar mills). What remains in place is state control over purchasing and inputs as well as disposal of land and main crop selection (Fernandez et al. 2018; Royce 2018).

At the same time, tillage and agrochemical application has fallen sharply since the 1990s, affecting only a tenth of farmland, including in private holdings. There are also some negative trends that constitute another front of struggle. A parallel market based on US dollars and a rising private sector (not only in farming) are contributing to rising social inequalities that are not sufficiently counterbalanced with more egalitarian wealth redistribution mechanisms, such as cooperative farms (Levins 2005; Rosset and Benjamin 1994).

As remarked above, the turn to ecologically sustainable farming was accomplished with much state and scientific community support in the diffusion of agroecological principles and the bolstering of traditional methods (Fernandez et al. 2018). The conversion has not necessarily been smooth, as tensions have occurred with farmers in some instances and there remain divisions between those promoting conventional and agroecological techniques (Fernandez et al. 2018; Royce 2018). The ANAP-based farmer-to-farmer movement in part may be facilitating the resolution of such tensions by demonstrating higher productivity rates through agroecological methods, gradually pulling the rug from under die-hard supporters of conventional agriculture.

Agroecological production methods are devised to maximise farmer autonomy by relying as much as feasible on local ecosystem resources and recycling them within the same locality. Techniques include biological pest removal, refraining from agrochemical inputs, manual weeding, nonhuman animal traction and minimal tillage, raising biodiversity and conserving water through polycultural growing techniques (e.g. intercropping), crop–livestock integration, mulching, seed saving and maintaining and raising soil organic matter levels through such techniques as recycling biomass into soils through composting and vermicomposting. Institutional adoption of intercropping techniques for small-scale farming has acknowledged the importance of and built on combinations of methods developed by Taino-Arawak and West and Central African farmers. This way, existing traditional knowledge has been valorised and recovered, while science has been redirected towards more positive ends (Fernandez et al. 2018; Machín Sosa et al. 2013; Rosset et al. 2011; Rosset and Benjamin 1994).

The push for agroecological methods was formalised in 1987 with the founding of the Cuban Association of Agronomists and Foresters, whose charter is the promulgation of ecologically sustainable farming and silviculture by means of research, training and extension service. Decades of efforts bore fruit by the 1990s when agroecology found major support in ANAP, within which, in 1997, there emerged a largely volunteer-based agroecology movement. It is based on horizontal communication and mutual aid among farmers (farmer to farmer), in contrast to unidirectional 'expert' extension services. ANAP started promoting low-input agricultural techniques and became members of La Via Campesina, an international small farmers association founded in 1993 to promote food sovereignty among other social justice objectives. This way ANAP members could even more easily meet, exchange ideas and learn from farmers in other countries in the region regarding agroecological techniques (Machín Sosa et al. 2013).

The development and diffusion of the farmer-to-farmer programme sped up the diffusion of ecologically sustainable farming methods. The farmer-to-farmer movement focuses on integrating traditional farmer knowledge and farmer innovation with agroecological science through information exchange programmes, workshops and free training, which are also part of ANAP's remit since its founding. As a grassroots initiative involving 216 households in 1999, the farmer-to-farmer movement grew exponentially to include 110,000 farming households by 2009. In 2018 the number has risen to 200,000 (Lucantoni 2020).

Aside from institutional backing and Via Campesina facilitation, the agroecology farmer-to-farmer movement received funds from international donors like Oxfam, helping the movement thrive independently of the state. Ecologically sustainable practices were thereby developed and spread throughout Cuba. Among the palpable results is the increase in maize and bean yields achieved with a net reduction of agrochemical fertiliser application at the national scale (Betancourt 2020). The extent of famers' adoption and eventual conversion to agroecological techniques and their ecologically constructive effects are not detectable under 'organic farming' figures, used by some to analyse progress made in ecologically sustainable farming. In Cuba only 0.2 per cent of farmland can be considered under 'organic' management, amounting to about 15,500 ha spread over 7,100 farms (Willer et al. 2008, 233). It must also be borne in mind that 'organic' certification exists mainly to fulfil 'organic food' export interests. In fact, US businesses have been eyeing Cuban organic farms as potential cheap suppliers to a burgeoning US organic food market (Severson 2016; see also Fernandez et al. 2018, 15–17).

The beneficial social and ecological effects of farms' adoption of agroecological techniques are excluded by such labelling, especially if some agrochemicals continue to be used during transition phases. Furthermore, agroecological principles emphasise adapting to local ecological conditions (rather than simply substituting agrochemicals with alternative 'organic' substances) and maximising farmer autonomy (rather than reproducing dependence on capitalist markets for inputs) (Rosset et al. 2011, 163–4). A more relevant datum would then be that cooperatives of smallholder farmers produce 65 per cent of Cuba's food on a quarter of total farmland and produce more food per hectare than large, industrialised farms. Between 46 and 72 per cent of these farmers (numbers vary depending on the province) apply agroecological techniques to differing degrees, even for cocoa, maize, rice and tobacco cultivation and milk and meat production (Altieri et al. 2012). It will take time for a transition to solely ecologically sustainable farming, including eclipsing import dependence for such staples as cereal grains, but in Cuba, as nowhere else, the transition is in full swing.

The Rise of Urban Farming in Cuba

Among the most impressive shifts in Cuban agriculture has been in the cities, where urban farming has taken off to levels unimaginable a few decades ago. This is a crucial turn because by the 1980s the number of people living in cities reached between 69 and 80 per cent of the total population (Marshalek 2017; Rosset and Benjamin 1994). Millions of tonnes of vegetables are now produced agroecologically over more than 50,000 hectares of urban land by 383,000 urban growers. There is no other country where low-input, ecologically sustainable urban food production has reached this degree of success (Altieri et al. 2012).

The case of Havana well exemplifies the transition process and the biophysical transformations involved. There, the first urban farm (*organopónicos*) was set up for civilian uses in 1991 (Koont 2011, 25). Food production long existed in Havana, but was institutionally suppressed and relegated to backyards after the 1959 revolution, until the late 1980s revival. The city, currently inhabited by about 2.1 million people over roughly 782 km², is a major port and commercial centre over karst topography and traversed by eleven short, low-flowing streams and the 50 km long Almendares (Febles-González et al. 2012), which has been contaminated with heavy metals (e.g. cobalt, chromium, lead) by an upstream smelter and landfill (Olivares-Rieumont et al. 2005). The climate is tropical savanna with marked seasonality.

However, the main meteorological influence (and hazard) comes from droughts and hurricanes, ever more frequent and intense, as well as increasing sea levels. The state-led introduction and spread of agroecological techniques, however, has not only been transforming urban ecosystems by raising agrodiversity, reducing exposed soil surfaces and decreasing agrochemical contamination, it also has resulted in the improved resilience of food-producing areas to the onslaught of more frequent and higher-magnitude extreme weather events related to global warming (Altieri and Funes-Monzone 2012).

High-intensity rainfall can still cause much flooding, due to insufficient draining and channelling capacity. Groundwater and aquifers are also increasingly affected by salt-water intrusion resulting from sea-level rise (Placeres et al. 2011).

Most soils are red and chalk brown earths, moderately alkaline and typically with sufficient nutrient levels for fruit trees, pastures and cane sugar. Towards the coastal areas there are cases of exposed karst (Placeres et al. 2011). In some cases soils have been contaminated through open solid waste dumping over the past century (Rizo et al. 2012). There are also sources of organic pollutants, like an oil refinery and smelter. A recent study showed that contamination levels for PCBs and PAHs are largely within safe concentrations in most urban soils sampled, except a few city parks. The matter is of concern relative to direct soil particle contact or inhalation, but not food production (Sosa et al. 2019).

Heavy metals, though, may be problematic. Lead and zinc are particularly high in industrial zones, but cobalt and nickel are in part geogenic (from the rock or sediment layers below soils). Industrial plants and other installations amount to 197 point sources of pollution and diffuse sources mostly associated with motorised vehicles, though cases of smog tend to be rare and particulate matter levels are usually low (Placeres et al. 2011). Lead and other trace element contamination is a concern downwind and near to industrial point sources, while contamination via vehicular traffic seems as yet understudied (Álvarez et al. 2017). Aside from industrial areas there are school grounds and city parks with an incidence of high levels of cobalt, nickel, zinc and copper from human sources, which may be due to aerial deposition from industrial and vehicular emissions (Rizo et al. 2011). Given the prevailing alkaline conditions of soils and the widespread use of compost in urban cultivation, the main contamination threats are likely from airborne sources and possibly through watering and storm-related runoff. This is because the potential for vegetable contamination by the heavy metals listed is negligible with high pH soil and high levels of soil organic matter.

Urban cultivators have had to face these and other biophysical processes and will be facing possibly greater difficulties with the effects of climate change. Among the more salient ones is the scarcity of irrigation water amenable to crop production (Koont 2011, 180). Salt-water intrusion into aquifers and the contamination of the Almendares River exacerbate this problem. Urban food producers, on the other hand, have been bringing about major ecological changes. One is the forgoing of agrochemicals and farming machinery in favour of organic farming techniques and agroecological applications. Such changes include overhauls in land use and land access distribution. Another is in contributing to urban reforestation (including fruit trees) and green space expansion more broadly, as urban food production units also participate in a national greening programme (Koont 2011, 175–6). Urban heat island effects and pollutant dispersal can be radically reduced in this way, while moisture can be retained more effectively and flooding magnitude could also be mitigated. Agrodiversity, if not total biodiversity, may also be increasing as a result of the spread of ecologically sustainable farming.

The blossoming of urban farming by the 1990s is the fruit of heavy state promotion combined with initiatives and pressures from below (French et al. 2010; Marshalek 2017). The virtual disappearance of small, private gardening and animal husbandry by the 1960s gave way to greater dependence on rural farming. By the late 1980s, under duress from a continuous US embargo and a major shift in conjuncture, a sea change occurred. The government was to some extent compelled to concentrate on urban food production because most Cubans live in cities. The now celebrated, larger and more commercially oriented urban farms have been supplemented by smaller usufruct-based *parcelas* on public land and home patios (now enjoying official appreciation) that provide for subsistence as well as private earnings (Altieri et al. 1999; Koont 2011, 165; Levins 2005; Machado 2017; Rosset and Benjamin 1994).

Urban food production in Havana and in Cuba is exceptional. Not only has it been reintroduced and supported by state institutions, but it is also integrated into wider agricultural planning, including peri-urban areas. This level of coordination is possible when the national state retains tenure over most land, private enterprise is restricted and profitability is subordinated to a primary directive of feeding people. Furthermore, the policies of the Cuban government over the previous decades have been crucial to establishing the research and extension structures, educational levels and skilling processes that became important in confronting the sudden economic downturn of the 1990s and promoting the development and implementation of technical innovations and improvements for urban farming. This is mainly through the above-described farmer-to-farmer movement coupled with specialised scientific institutes diffusing agroecological applications through ANAP. State institutions have played key roles in providing the inputs, material incentives and moral inducements (e.g. patriotism rhetoric) to diffuse agroecological and organic farming methods (Koont 2011, 8; Premat 2012).

Estimates on the number of urban cultivators are contradictory. Some claim 50,000 and others 90,000 people are involved in urban food production. Regardless, this is a large number of people who have become involved and it is especially noteworthy that 7 per cent of Cuba's workers are formally employed in urban agriculture (Koont 2011, 191). Some of the reasons lie in economic benefits potentially gained by cultivators. Roughly 40 per cent of Havana's food production includes selling surplus (González Novo and Castellanos Quintero 2014). The end result of combinations of inducements is that vegetable and fruit production has reached levels hovering at or exceeding minimum levels to cover the city's nutritional needs. Much of what is grown, especially on the *organopónicos* and intensive gardens, covers everyday popular culinary needs and abides by agroecological principles, also diffused via state extension programmes and farmers' own innovations (Leitgeb et al. 2011). Beans, gourds, lettuce, melons, plantains, tomatoes and watermelons, for instance, are grown using composts and more concentrated organic fertilisers and in combination with herbs and other plants helpful in warding off pests (French et al. 2010, 158; Leitgeb et al. 2016).

Prospects for Cuban Urban Agriculture

Under current circumstances (international as well as national), urban cultivation can complement food production of staple crops in the countryside, where organic techniques have also been diffused institutionally. This way, shortfalls in food production and access can be overcome more easily at the same time that there emerge and develop greater and more diffuse self-reliance in society and more ecologically sensible practices. The fate of urban cultivation is said by some to hinge on the prevailing economic trajectory, which is also linked to modifications to US policies and wider capitalist world-system shifts. The recent rapid decline in larger gardens (including *organopónicos*) was in part to give way to more lucrative tourism and manufacturing use (French et al. 2010, 159). This appears to indicate an inverse relationship between economic hardship and urban food production similar to many cities in the Global North. However, the continuous rise in the numbers of smaller *parcelas* and patios (Koont 2009) contravenes any notion of overall urban food production decline and instead points to policies and movements from below that support the spread of more decentralised practices, as well as displacement to peri-urban areas to cope with economic tendencies (from within as well from abroad) privileging activities that generate higher earnings, attract foreign investments and enable survival in the face of relentless US imperialist pressures.

Productivity may be relatively high, especially when compared with gardening in cities over the Global North, but prospects may be tied to what people decide to do as the harshness of the 1990s is giving way to improved living conditions and possibly greater availability of fossil fuels and agrochemicals (Koont 2011). Urban food growing and farming generally have been insufficient to ensure food availability nationally. Consequently, Cuba continues to depend on imported foods (up to 40 per cent of total food consumed, if not more). Much is being made of this difficulty as part of an effort to show the alleged failures of the socialist state in Cuba (Bruno 2017; Machado 2017; Montgomery 2007). This point of view ignores the larger picture. For an island country hit by a decades-long embargo it is a herculean and long-term effort to exit import dependence and general colonial plantation legacies, reinforced through decades of trade with the USSR bloc. It is unreasonable to expect Cubans to achieve what no other country in the Caribbean (and beyond) has yet to achieve (Levins 2004). For example, another island nation, the UK, imports 50 per cent of its food (DEFRA 2017).

The difficulty with fulfilling food demands in Cuba may also be that much of the imported food is made out of crops that cannot be adequately grown under tropical island conditions, such as cereals and soybeans, without resorting to fossil fuel-derived agrochemicals. Dietary patterns are also tough to change when the prerogative is ensuring all Cubans' nutritional needs are met. This is a prerogative that, for instance, does not exist in the US, where hunger persists amid one of the highest food production levels in the world. One could also argue that an overwhelmingly urban society needs to undergo major change over a couple of generations to transform itself into an agrarian society that uses fossil fuels sparingly. What should be taken as much more consequential is that the benefits of urban farming have not been evenly spread in Cuba. There persist social justice issues due to racialised class disparities, unresolved since before the 1959 revolution, that hinder the attainment of equal food access for Afro-Cubans (Lowell and Law 2017, 112–13).

There are still other international and contextual aspects that should be considered. Part of the fate of urban food production and the ecosystemic impacts it implies is tied to newer linkages being forged since the early 1990s. One of them is with the PRC. Its capitalist reorientation notwithstanding, the PRC have played an important, if underappreciated, role in the transfer of technical know how and materials helpful towards the development of low-input, organic farming methods (INIFAT 2010, 11, 18). Moreover, there are cultural dynamics within Cuban society that have also in some ways facilitated the urban cultivation renaissance, aside from internationally derived pressures. Some of the early and persisting forms of highly intensive urban cultivation over inhospitable, tiny spaces are traceable to the gardening practices within Cuba's long-standing Chinese diasporic community (see also Koont 2011, 182). Perhaps, then, it is not so random that General Sio Wong, the revolutionary war veteran of Sino-Cuban background and party leader, has been a main proponent of urban farming and agroecological principles (Koont 2011, 25–6; Premat 2012, 11). One urban cultivation method, promoted by institutional experts and the main basis of *organopónicos*, is to mix sediment and composted materials to help establish growing areas over virtually uncultivable city surfaces. These are also part of traditional expertise and everyday gardening practices.

The above major achievements are disparaged by detractors as the effects of economic crisis, the US embargo and a partial opening to the 'free market' (e.g. Benz 2020; Montgomery 2007). These are the sorts of specious arguments that ignore how similar embargoes or economic hardship in other countries, such as Iraq, have been met by privileging military or other exclusive, narrow interests over safeguarding most people's well-being (Cabello et al. 2012). Moreover, the implication that only economic downturns can bring ecologically beneficial effects is politically rather sinister, implying that social struggle and political change are irrelevant. Unlike those social systems, the state socialism of Cuba, where social equality and well-being are priorities, is the sole existing model that meets a minimum requirement of standards of living 'within a consumption pattern that could be extended globally without entering [planetary] overshoot' (Moran et al. 2008, 4). That is no small feat for a small country under siege by the mightiest military in world history.

THE NET ENVIRONMENTAL EFFECTS
OF STATE-SOCIALIST SYSTEMS

For the most part, especially in Cuba, the net ecological consequences of state socialism have been an improvement over pre-existing conditions under the authoritarian rule of rapacious capitalist (and associated semi-feudal) institutions, well networked with or under the thumb of the colonial dictatorship of liberal democracies. If some impacts of socialist states were negative, in some instances for the long term, it is not for a lack of ecological sensitivity within ruling parties or the wider socialist sections of state-socialist systems. It is in part a result of the eventual dominance of ruling party fractions and sections of society who privileged industrialisation at any cost and the shifting struggles within socialist states over priorities. Just as important, though, are the external inimical and belligerent forces and historical circumstances beyond the control of socialist states.

Environmentalism was constitutive of, not foreign to, the socialist state, and even more so with Cuba and the USSR at large. There was no need to import or smuggle environmentalist work into a country with a long history and wealth of ecological thought, especially in the case of the USSR (Elie 2015). Before the publication of Rachel Carson's germinal *Silent Spring*, the biologist and environmentalist Aleksandr Nikolaevich Formozov was calling attention, in 1961, to the devastating effects of pesticides on birds, insects and other organisms, including those beneficial to agriculture (Weiner 1999, 305). Under state-socialist systems some of the most advanced environmental research and monitoring programmes were developed. The priority given to mass literacy and scientific education, coupled with campaigns of ecological sensibilisation by scientific societies, contributed to public sensitisation to ecological issues and to successful environmental struggles. This remains the case in Cuba.

The challenge was and has been throughout state-socialist history reconciling industrialisation with environmental protection, which manifested itself by way of struggles among and within ministries as well as wider activism outside of government, involving scientific communities and volunteer organisations founded by socialist currents during the early revolutionary periods. The unresolved conflict between pollution and industrialisation reflects a contradiction between achieving socialism, which meant promoting the improvement of everyone's well-being as well as nature protection, and developing the material means to do so. The net ecological effects (and thereby also the effects on people's health) were therefore mixed. There were major improvements in nature protection and conservation, enhancing biodiversity, greening cities and reducing some forms of air and water pollution over time. And there were major pollution problems from industrial production, including farming, along with several localised and regional instances of lasting destruction. This was certainly so for the USSR and the PRC, but much less so for Cuba.

That such a challenge between aims and available means remained and remains unresolved is to some extent traceable to the ways in which decision-making processes were structured. For instance, in the USSR people running mining, energy and other industrial production sectors were rewarded with bonuses for meeting production targets in a timely fashion. Anything that impeded resource provision or that caused delays to those ends were strenuously fought against and environmental and health effects were dumped on the central state organs to resolve.

There also were and are, in Cuba, rewards for nature protection and public health actions, in terms of promotions, research funds and the like. And then there are the thousands of environmental activists within and outside formal institutions whose reward lay mainly in preventing harm, reducing impact or pushing the state to enforce environmental legislation. This is in addition to the capacity of state-socialist governments to mobilise, at times coercively, at times via material incentives, millions of people to do ecologically constructive things. This is exemplified by reforestation efforts in all the countries discussed, monitoring from below in the USSR and Cuba, or vermicomposting training workshops in Cuba.

The selection, allocation and use of resources are part of an overtly political decision-making process in state socialism, not just a technical matter or an issue decided within private companies without any public input until after the fact. The centralised decision-making system and planning arrangement is both an advantage and disadvantage. On the one hand, there are clear ultimate responsibilities over plans, regulations and actions relative to the environment and ministries could be played against each other to improve overall results. On the other hand, because central authorities' legitimacy rode on providing well-being to all, including liveable environments and nature protection, any discrepancy between policy and results immediately undermines the state politically. Even environmental information flow is part of overt, if not formalised political struggle.

The outcomes of these struggles have differed. In the USSR and the PRC the central authorities (such as the Politburo), rather than concede their major faults and change course, opted to conceal and repress as much as possible, fearing a backlash, which happened multiple times and in multiple places regardless. Air and water pollution cannot really be hidden from the rest of the citizenry. Such instances of environmental damage, in a state-centralised decision-making context, become crises of political legitimacy. In Cuba the matter was resolved through greater decentralisation and allowance of public criticism by the 1980s (Levins 1990). The question regarding the USSR or PRC should be why the central authorities made such (in the long-term self-destructive) decisions compared to countries like Cuba.

Progress on environmental issues in a country like the USSR or the PRC under Máo could have been made to reduce if not prevent the major air and water pollution coming out of the industrial sources that experts had identified if the USSR or PRC had amassed the level of capital and material extraction to accomplish that. But that would have necessitated drawing resources from most of the rest of the world by imperial and economically exploitative means, as in the case of western European countries or the US and Canada. The exploitative route is essentially alien to socialist states and is in any case unfeasible. To do that takes many more decades of conquering and ransacking other countries and then maintaining a neocolonial relationship.

Hence the USSR, like China until recently, developed the technical capacities to prevent or overcome much ecological damage associated with rapid industrialisation, but without the economic means to reach that kind of technological level. It was then a choice of further exploiting domestic resources and workers, as China has done, or borrowing and importing technology and spare parts from the capital rich countries, as most states do, and thereby become indebted to governments that endeavoured to destroy any form of socialism. The USSR leadership opted for greater economic dependency on the core capitalist countries, managing to resolve neither the economic and technological independence problem, nor avoid some major environmental problems.

In Cuba, however, matters were taken in a somewhat different direction, seizing the imposition of an economic disaster as an opportunity to build an ecologically sustainable socialist society. This is in itself unprecedented and squarely associated with the development of a state-socialist system. The many capitalist countries recently experiencing sudden and devastating economic downturns, like East Asian countries in 1997 or European and North American countries in 2008, did not result in any serious institutional rethinking (certainly no self-critical analysis) or substantive efforts of the sort that the Cuban socialist state has undergone. This should not surprise anyone. Since the prime directive is profitability and wealth accumulation, capitalist institutions are structurally incapable of mobilising society in a constructive direction. It takes socialist revolution or massive organising and pressures from below to push against the undemocratic nature of capitalism and reach the chance to build an egalitarian and ecologically sustainable society.

Cuba thus presents an acute contrast to other countries on account of advances in ecologically sustainable practices since the revolution. Forests expanded along with biodiversity, ecological preserve areas have been multiplied and enlarged, and farming has increasingly shifted to ecologically sustainable techniques that are so advanced as to be the envy of the world. What is more, food production has been integrated into urban centres to a degree unmatched anywhere else and affording large outlays of produce in ways that improve urban ecosystems. The country is exemplary in implementing ecologically sensible measures and substantively reducing environmental impact, while markedly improving the well-being of society as a whole. While enforcement and shifting to renewable energy are a challenge when having meagre means, the efforts have borne fruit if the ecological footprint is any guide (Figure 4.19). Cuba was consistently in the lower portion of ecological impact among countries with similar values for the UN's Human Development Index (HDI), but by the 1990s it was consistently the least ecologically damaging with half as much impact as Costa Rica, the closest country in the same HDI range (Cabello et al. 2012). The Cuban government's achievements in conservation are acknowledged even in the *New York Times* (Goode 2015), where, daringly, it is intimated that a resumption of US business activity on the island will bring the environmental woes of yore (see Fernandez et al. 2018, 18).

5

Reckoning with Contradictions to Build Ecosocialism

Sweeping condemnations of the environmental impacts of socialism can and should be easily swept aside. They are patently baseless for several reasons. The more obvious one is that the wide historical diversity of socialist currents (which here include anarchist and communist ones) is not reducible to state-socialist systems. Only some of them are associated with the establishment of socialist states. Second, a more comprehensive environmental record shows that the net impacts of socialist states have been positive. This is on account of ecosystem and soil conservation, reforestation and afforestation, at least partially successful measures in air and water pollution abatement and prevention, and, most of all, relatively low ecological footprints. Another reason is that many socialist states did not make much of a dent on ecosystems relative to countries with greater historically accumulated capital. This is in part because most socialist states did not industrialise or had no chance to (see Rodney 2011). Studies on the environmental impacts of 'communism' overwhelmingly exclude most socialist states without any justification and this raises the suspicion that such largely Eurocentric scholarship is underlain by a white western European superiority complex. A fourth reason is that the environmental record, as also shown in the previous chapters, has been much worse for capitalist countries and keeps getting worse over time, to the point of having already caused planetary, not just localised or regional, calamities. Finally, the more egregious cases of environmental destruction caused through socialist states bear at least in part a direct and indirect imprint of capitalist countries, especially liberal democracies, by means of historically inherited environmental devastation, sustained military and economic pressures and exacerbation of environmental degradation following state socialism.

Deleterious and disastrous long-term biophysical impacts, such as there were, resulted in some socialist states, mainly those undergoing industrialisation. Those are what require explanation. To some degree an explanatory framework has been hinted at or expressed throughout this work by means of analysing the wider context in Chapter 3 and concrete situations in Chapter 4. Allusions have already been made to the contexts wherein socialist states arose and developed. An explanation is offered later in this chapter that is based on a historical and dialectical materialist framework. Afterwards, the framework is extended to consider some overarching questions and quandaries compelled by the historical challenges faced by socialist states. An argument is made that reclaiming state socialism is important towards exploring realistic ecosocialist possibilities in the present. But first, here is a short overview of prevailing explanations about state socialism and the environment.

PRINCIPAL NARRATIVES ON STATE SOCIALISM AND THE ENVIRONMENT

Three overarching and interlocking stories pervade the literature on state socialism and the environment. One focuses on alleged state-socialist residuals or nasty legacies. Another view traces most or all problems to the authoritarianism or 'totalitarianism' of state socialism. A third line of reasoning would have Marxist ideology as the culprit. These explanatory approaches rest on capitalist democracy supremacism and presume a mainly or totally negative state-socialist environmental record. For these reasons alone they are irrelevant because of the mostly constructive effects of state socialism, as shown above, and because of a long record of environmental devastation in capitalist societies. The conventional storylines could apply to cases of state-socialist environmental degradation, though, if they did not fail on their own terms or through contradictory evidence.

The state-socialist environmental legacies argument is downright embarrassing. Basically, an alleged disastrous present is the fault of what transpired in the past under state socialism (e.g. Agyeman and

Ogneva-Himmelberger 2009; Pavlínek and Pickles 2000; Peterson 1993). Is the pollution of Lake Okeechobee or the Mississippi Delta Dead Zone then a legacy of liberal democracy? The narrative unfolds as a selective chronology of inheritance, where the past conveniently stops at the beginning of a state-socialist system. Environmental pollution or degradation preceding socialist states are underestimated or eschewed, as are constructive legacies from socialist states. Worse, the legacies argument can also serve to absolve institutional responsibilities for environmental harm after socialist states were dismantled.

A second explanatory strategy resorts to the tired 'totalitarianism' construct or, in less ossified renditions, finds the ultimate causes in authoritarianism and related institutionalised conflicts of interest, including a lack of checks and balances, secrecy and censorship (e.g. Dominick 1998; Edelstein 2012, 38; Goldman 1972, 34, 138–40; Goldman 1992, 3–4, 214; Josephson et al. 2013, 13; Mignon Kirchoff and McNeill 2019, 13; Pavlínek and Pickles 2000, 6, 26; Peterson 1993, 12–18; Pryde 1972, 2). Authoritarianism, in that perspective, usually means state or central party organ control, lack of a 'free market' and market signals, and private property, among other hallmarks of capitalist democracy (e.g. Goldman 1972, 40, 46–8; Josephson et al. 2013, 207–9; Pavlínek and Pickles 2000, 16; Peterson 1993, 13; Pryde 1972, 42, 119–20; Weiner and Brooke 2018, 309).

At a most banal level this is not an explanation at all for state-socialist malfeasance because similar or worse records exist under free-market democracies. Furthermore, the existence and frequent effectiveness of environmentalist opposition, reported by many of the same scholars, contradicts the claims made on their own terms. Another fly in the ointment is the institutional diversity existing within socialist states and the antagonisms among such institutions, involving wrangling over state funds (Goldman 1972, 169; Jancar 1987, 5). Additionally, natural resources and means of production were often under cooperative management and to some extent under smallholder private ownership, not direct state control. The notion of ultimate Politburo control is largely fictional. Finally, the claim of censorship and secrecy is spurious, as detractors of state socialism themselves describe many instances of substantial data access and of open criticism without repression (Josephson et al. 2013, 150, 157; Goldman 1972, 34, 54, 65–6; M.I. Goldman's foreword in Komarov 1980; Obertreis 2018; Pryde 1972, 68), while overlooking copious cases of environmental information suppression in liberal democracies. There were even cases of collusion by the US government and the nuclear industry when it came to suppressing information on the Kyshtym and Chernobyl nuclear accidents (Brown 2013; Josephson et al. 2013, 259).

A third major line of reasoning attributes causal powers to ideas and faults Marxism for environmental degradation (e.g. Jancar 1987, 6; Pryde 1972, 42; Shapiro 2001, 4). Again, given environmental devastation in liberal democracies this is not a credible explanation. In any case, there are many kinds of ideas and belief systems that eventuate in the same environmental effects (cf. Marks 2017). That there have been different and even mutually antagonistic Marxist currents also seems beyond such theorists' comprehension.

None of the arguments offered in the mainstream about the relationship of state socialism to the environment address why socialist states scored so many environmental successes. They cannot even explain what they are fixated with, which are instances of environmental degradation and devastation in the industrialised state-socialist countries. The solutions they offer have at best led to mixed environmental outcomes, if not more destruction, in the very liberal democratic countries used implicitly or explicitly as comparative paragons. What is sorely missing in studies of the environmental impacts of socialist states is not only a balanced framework, but also social and ecological contextualisation. For example, deforestation is impossible to explain by looking only at where the deforestation happens. This is a basic methodological principle developed through several approaches over five decades, such as political ecology and the world systems paradigm.

CONVERGENCE AS EXPLANATION

Frameworks that steer clear of crude liberal democratic supremacism have recently surfaced under the 'revisionism' label. They are attentive to the wider context and to historical open-endedness, overwhelmingly about the USSR, and are more mindful of capitalist horrors. However, the claim is that environmental impacts in capitalism and state socialism tend to converge in level of harm or forms of environmental practices. This line of argument resembles the more moderated versions of the views expressed by the likes of Goldman (1972) and Pryde (1972). Where convergence proponents differ is in concluding that state socialism (or 'communism' as they often frame it) is not in itself worse for the environment. They recognise factors beyond the control of socialist states, including biophysical ones, and underlying social structures shared by capitalist systems. Where they differ is about the source of convergence. Some find it in 'the common experience of modernizing states', premised on a 'growth imperative' (Bruno 2016, 15–16), 'developmentalism' (Pomeranz 2009a; Weiner and Brooke 2018, 309) or Enlightenment ideology (Josephson et al. 2013). Some recognise the role of global economic pressures alongside questionable technical decisions (Bruno 2016, 20) and see deep-seated continuities with the tribute-seeking statecraft from centuries ago (Weiner 2009). It is all part of a general worldwide trend in spiralling population growth, energy use and economic growth that took off in the 1970s and 1980s (Chu 2018, 158). In other words, the explanation is really an observation that state socialism converged with liberal democracies and the rest of the world when it comes to environmental impacts and only painstaking comparisons can reveal differences in the causal mechanisms that led to such convergent patterns.

Convergence perspectives, though, cannot explain the general global trend and the timing of the worst effects in the 1970s and 1980s, nor why environmental destruction has intensified since the demise of the USSR. It also leaves unaddressed whether and how the same problem extends to all state-socialist countries. This is crucial. The Congolese and Mozambican socialist states did not undergo industrialisation, while the Hungarian and North Korean ones did. Cuba's environmentally constructive record also remains inexplicable on this view. Revisionists cannot specify the overarching social processes whose outcomes are ecologically unsustainable or spell out and develop political alternatives. It is a viewpoint that sidelines questions of relations of power at multiple scales, including settler colonialism, neocolonialism and imperialism. Treating socialist states as analogues of 'modernity' is a politically evasive and disabling manoeuvre, dissembling relational differentiation and major unevenness in power relations, while collapsing widely differing social and historical conditions into an undifferentiated 'modernity' porridge.

The emptiness of this kind of perspective is well represented by a question posed by some 'revisionists' on account of the discomfort associated with realising the equally terrible natures of capitalism and 'socialism': 'Is it possible to create an economy that respects society, democracy, and nature?' (Mignon Kirchhof and McNeill 2019, 14). Their discomfort is understandable, but it is predicated on asserting equifinality without considering the evidence, as presented here, that state socialism has more constructive environmental effects than capitalist democracy. Since they believe all state socialism, without bothering to study its multiple forms, is little different from liberal democracy (the revisionists' fixed reference point), the conclusion is that environmentally friendly impacts are independent of political system. It makes no difference whether there is democracy or dictatorship.

And this is true. Dictatorship is quite compatible with reducing environmental damage, if not even preventing it. Authoritarianism has been standard practice in setting up national parks in settler colonial systems like the US and Canada, for example, forcing Indigenous peoples out or killing them. The reason why the notion of a just transition or environmental or ecological justice is crucially important is precisely because there is no necessary link between egalitarian social relations and constructive environmental impact. Ecosystems and biomes are much wider processes than what happens within a single species. The political ambiguity of environmental issues makes it all the more important to be transparent about political commitments. Revisionists suffer from the same political ailment of third-way environmentalists who claim both socialism and capitalism are destructive and then end up siding with whoever is in power to achieve effective environmental protection measures.

EXPLANATIONS FROM LEFTIST POSITIONS

There is another interesting twist to the stories recounted about the alleged failure of socialist states on the environmental front. Goldman (1972, 64–6) was among the few in the mainstream who showed a grasp of how pressures to overcome 'underdevelopment' led to environmental trouble. Even if claiming that socialist states are environmentally damaging is an unsupported assertion, this argument by Goldman is interesting because of its coherence with the directives (or justifications) from state-socialist governments. It also highlights one of the central concerns of socialists more broadly relative to improving everyone's quality of life instead of that of a select few, as in capitalist conditions. This already creates major challenges when, as already pointed out, a country is in a state of siege and is under economic pressure, not to mention divergent internal factions struggling over development and survival strategies, which entails struggles for power. There was no outlet to displace the polluting consequences of industrialisation, such as far-flung colonies or advantageous neocolonial relations and terms of trade favouring outsourcing and relocation of environmentally destructive economic activities (O'Connor 1998; Weiner 2017).

Several explanatory frameworks have already been developed on the left, but they presume that state socialism is environmentally wretched. Because of this they are compelled to construct a narrative that distances state socialism from their brand of socialism. Hence, they trace the causal mechanism in the repression of dissenters, including many ecologists, and the suppression and distortions of Marxian text, combined with external pressures, eventuating in the concentration of power and class neodifferentiation, authoritarian economic planning and the prioritisation of manufacturing, among other consequences leading to environmental harms (Angus 2016, 208–11; Foster 2000; Gare 1993; O'Connor 1991, 9; Sheehan 1985; Weiner 1988). A still underexplored avenue is the major role in environmental destruction played by persisting patriarchal relations and the militarised government that derives from them. These have been hitherto only briefly tackled (Seager 1993, 111–16). Gare (2016, 21) finds that the recent neoliberal 'fusion of bureaucracies and markets against democracy' (and the striving for global control) coheres with the technocratic managerialism of socialist states, which have therefore had little difficulty in embracing this latest capitalist form. These tendencies are interrelated. As pointed out already, from the beginning socialist states faced enormous external pressures from liberal democracies in the form of constant military and economic pressures (Democratic Socialist Party 1999, 64–5; Chase-Dunn 1982; O'Connor 1998). The precision of these explanation are not in dispute here; the problem is that they have been missing the mark.

A system prioritising the overcoming of low levels of material well-being and productivity levels, and usually coming out of peasant patriarchal societies and often also colonial oppression (among other major disadvantages), all the while being militarily invaded and/or besieged, creates conditions not exactly conducive to the development of egalitarian social structures and environmentally constructive impacts. In these historical conditions it is a marvel that there were any successes in improving standards of living, raising environmental standards, engaging in reforestation, reducing or preventing waste, pre-empting consumerism, developing recycling programmes and attaining high volumes of recycling, improving access to education, providing free and universal health care, guaranteeing full employment and job security and even setting up lasting conservation areas, among other major feats. Add to this the development of cutting-edge social practices and policies like gender parity efforts in Burkina Faso (Biney 2018), reconciliation and pacification methods (even if fraught with violence) in Mozambique (Igreja 2010; Meneses 2016), revolutionary applications of agronomy in Guinea Bissau and the Cabo Verde Islands (César 2018; Manji and Fletcher 2013) and world-renowned successes in space, biomedical, health, conservation, engineering, ecology and biotechnology research and applications in the USSR, Cuba and the PRC (Fitz 2020; Foster 2015; Levins 2005; Navarro 1992; Weiner 1999), and the record is even more impressive.

DEVELOPING A DIALECTICAL HISTORICAL MATERIALIST
EXPLANATORY FRAMEWORK FOR ECOSOCIALIST ENDS

Another way to explain the relationship between socialist states and the environment is possible and necessary. One way to start is to look at ecosocial dynamics. In short, this means (1) a much wider historical and socio-geographical contextualisation than studies and explanations have so far encompassed; and (2) a dialectical materialist approach to biophysical dynamics that are inclusive of and beyond human impacts. I will stick to the negative impacts, since regrettably they overwhelmingly continue to catch leftists' imagination, but also because, given consistent historical patterns of capitalist violence and extreme pressures, the worse should be expected following a revolution.

For the first aspect, an example can be the historical and wider conditions through which socialist states developed. What has not gained enough attention in the copious literature is arguably of greater import than explaining the internal power struggles within the USSR and then in other socialist states that led to environmental disasters. The case of Chernobyl could be reconsidered this way. The existence of an atomic-nuclear programme in the USSR – first, in the early 1940s, for military and then, by the 1950s, for civilian purposes – resulted from a primary motivation called self-defence. Between 1932 and 1939, following the Japanese invasion of Manchuria, the USSR and Japan warred allegedly over boundaries, with tens of thousands of deaths. The USSR had already been invaded (from the east) by the US and their allies during the Russian Civil War, and the Japanese imperial expansion was possibly an even greater danger. During World War II the USSR was invaded by German armies assisted by various governments (e.g. Hungary, Romania) and nationalist contingents from eastern Europe (particularly Ukrainian). It was the Nazi government's largest military operation, it must be recalled (in practice, anti-communism trumps illiberalism in the Nazi mindset). The democratically elected US government bombed Hiroshima and Nagasaki in August 1945. NATO was founded in Washington, DC, in April 1949, to which the USSR and their allies responded only by 1955 with the establishment of the Warsaw Pact (in Warsaw, not Moscow, interestingly). It is also important to note that the USSR did not have any proven atomic bomb capacity until the first test in August 1949, months after the formation of NATO and several months before any certainty over the outcome of the Chinese Civil War (the PRC was not declared until 1 October 1949). By 1945 the US had encircled the USSR with military bases (Vine 2015) and had openly threatened the USSR with the atom bomb. It should be clear that the impetus behind the USSR's nuclear arms programme has been self-defence from the beginning. Unlike the US, on no occasion did the USSR threaten other states with nuclear attack. It should be self-evident that in a world heavily inflected by US belligerence one is almost compelled to develop a nuclear weapons complex. Otherwise, if a country becomes attractive to US ruling classes for certain resources or military strategies, the risk is to be invaded or bombed to smithereens, as happened with Vietnam, Laos, Afghanistan and Iraq, for example.

A related issue is why the USSR bothered at all to develop civilian applications of nuclear power. The potted answer is often high modernism, including a drive for industrialisation, developmentalism or productivism, but this suggests a sort of voluntarism (if not some innate Bolshevik evil, in some renditions of USSR history), as if people decide of their own accord to get into something awful or are somehow inexorably hungry for power and determined to keep it once they have it. What should instead be asked is what propelled such a turn. A similarly brief answer would be a combination of self-defence and getting legitimacy from most people by delivering on living standards improvements. Both aspects necessitate the building of higher productive capacity and with that comes some sort of centralised coordination and chains of command. None of this is inevitable and other roads could have been taken, but the immense military and economic pressure from liberal democracies must not be underestimated. In the USSR the first electricity-generating nuclear plant in the world used for a civilian power grid was set up in the USSR at Obninsk in 1954, about 115 km south-west of Moscow (this was followed in the UK, at Sellafield, and France, at Marcoule, in 1956, and then in the US in 1957, at Shippingport). This nuclear option was based on external military pressures and on finding the fastest route to the highest energy production, relative to technologies available.

This kind of reconsideration and reframing of the nuclear issue in the USSR is much deeper on relations of power than most perspectives offer. The example of Chernobyl is confined mainly to the scale of national states and does not address ecological processes, but one could easily go further to subnational struggles. Yet the question is, what would an ecosocialist formation do differently in the face of such conditions and in attempting to fulfil basic needs, while preventing environmental damage?

One way of addressing (not solving) the above quandary is to extend the contextualised analysis further by applying a dialectical materialist framework inclusive of biophysical dynamics, drawing from state-socialist achievements, and to learn from the environmental transformations under state socialism. Due to my prior work it is easier for me to discuss impacts on soils in Hungary, so I use that theme as an illustration. The example of impacts on soils in Hungary also coincides with the second aspect, delineated above, aimed at building a dialectical materialist framework.

In Hungary there were various transformations of soil use and science as a combined effect of the interlinkages of soils research, the outcomes of human impact and wider social changes at multiple scales. As farming became increasingly industrialised and central to the Hungarian economy by the late 1960s, under the Kádár government, soils were altered ever more destructively, encouraging major shifts in farming and soil science communities. Soil properties in some regions became more acid and/or overwhelmed with phosphorus and potassium. This appears to have been in mainly women-run household farming spaces more than conventional, large-scale plots (Engel-Di Mauro, 2003). Acidification through industrialised farming in humid temperate regions is often associated with prolonged and high amounts of nitrogen fertiliser applications combined with large-scale harvest removals and tillage (which leads to much soil organic matter rapidly degrading, while returns of organic matter to soils diminish). Other areas' soils with differing intrinsic properties remained near neutral or changed towards greater alkalinity, sometimes because of the amounts of fertiliser added were greater than ever before (this can raise alkalinity or neutralise acidity) and made available through industrialisation and centralised production systems. Such differential impacts on soils resulted from combinations of soil dynamics at variable timescales (millennia to years) and autonomous from society, as well as the outcomes of historical human impacts. The latter, even prior to state socialism, were traceable to Hungary's world-system position as net food exporter. Industrialised farming became especially important by the 1970s to develop agricultural exports to offset the Hungarian socialist state's dependence on finance and technologies from Western liberal democracies.

That there was environmental destruction, in this light, should scarcely raise any eyebrows. The relative damage to soils under the Kádár government is an instructive case, even as successful soil conservation also occurred and probably the most comprehensive and extensive soil-monitoring programme in the world was developed. Pressures to go in a negative direction were nevertheless great. Increasingly indebted financially to states and firms in Western liberal democracies, the Hungarian government introduced economic reforms promoting more privatised and greater export-oriented farming to garner the necessary currency to repay debts with interest from lenders in the wealthiest liberal democratic regimes (Berend 1996). These policies had the effect of intensifying mechanisation, pesticide and fertiliser production and application, and all the trappings of industrial farming that led to the now familiar path of soil degradation, such as compaction, erosion and acidification. And loan repayment along with internal pacification played an important part in such environmentally destructive outcomes. It was not then just a matter of accumulating capital through agriculture to facilitate the building or expansion of industry.

Changing soil properties and dependent centralised capital accumulation (e.g. financial indebtedness and agricultural exports to states and firms in western Europe and North America) brought about two contrasting and interrelated processes. On the one hand state soil science and agronomy gained a capacity for an unparalleled achievement in soil-monitoring resolution (to the scale of six hectares). On the other hand, the very capital accumulation that enabled that level of monitoring exposed the flaws of industrial farming and compelled scientists and cooperative farms to institute soil conservation measures and a liming programme by the mid-1980s. The realisation of widespread soil degradation problems, it must be borne in mind, would not have been possible without most farmland use being under centralised coordination. Such ecological and social transformations crucially involved the promotion of largely white, male and more profit-oriented interests while displacing women and racially minoritised farmers. By the 2000s those wealthier farmers who happened to grow crops on such previously limed fields would have some production advantages, especially compared to middle-income farmers who are squeezed by high productivity pressures while having insufficient capital (Engel-Di Mauro 2018). Finally, soil science itself has been deeply affected (even scarred) by all these developments, to which soil scientists themselves contributed actively. The socially shifting, but largely white male soil science community has contributed to changes in soils use while the outcomes of soil use have led to the reshaping of soil science communities towards even more entrenched productivism (in line with demands for raising agricultural exports, for example) or, sometimes as a form of political dissidence, greater environmental sensibility (Engel-Di Mauro, 2006).

More abstractly explained, these ecosocial processes can be viewed as intersecting and unfolding in a dialectical fashion. There are interpenetrating opposites (differentiations) that are part of a unity and at different scales (levels of organisation), such as limed and unamended soils, acidifying and neutral to alkaline soils, socialist states and liberal democracies, gendered and racialised farming communities, and industrialised and manual agricultural cultivation techniques. The interpenetration aspect can be exemplified by soil properties, preceding or autonomous from social dynamics, being both the subject and object of human impacts. Human impacts lead to alteration of soil properties, which change differentially beyond the influence of human impact, due to intrinsic qualities in soils. Altered soil properties influence how human impact unfolds, which in the above case raised a perceived need for higher fertiliser applications and even prompted a nationwide liming campaign, with repercussions on other ecosystems through quarrying, to name just one possible and wider ecological effect. In all this there are negations, such as the transformation of farming with industrialisation, export-reorientation and productivism to more conservationist forms of soil science and agronomy. Throughout, transformations have been mutual, between soil science and their subjects of study, policymaking and impacted environments, scientists and farmers, socialist and liberal democratic states, and so on. Most of all, there is no end to the story and no necessary predictability. The relationship between soil dynamics and social processes is material, dialectical and thereby open-ended. This could be another way of explaining what happened under state socialism in Hungary relative to impacts on soils, taking into consideration how human-induced transformations in soils contributed to changes in state-socialist policies. These brought about the potentials for a more constructive relationship to soils, but they were foiled by external pressures, like loan repayment subtended by cranking up farm exports, and internal contradictions. In this precarious situation the government attempted, among other things, to raise productivity to satisfy the objectives of improving people's living standards while at same time the party and wider society were internally divided among would-be pro-capitalist and unreconstructed Horthyist (with a few Nazis sprinkled in) counter-revolutionaries, factions bound to Moscow's priorities, and a minority of committed egalitarian factions.

One possible lesson from the above-described state-socialist ecosocial dynamics is the importance of anticipating and preparing for social struggles while developing the fullest grasp possible of biophysical processes, all at multiple scales of analysis (planetary–local). Relative to soils it would be crucial, before revolutionary prospects emerge and to make them happen, to organise a large cadre of scientists committed to agroecological and socialist principles while preparing to divide efforts between (1) developing environmentally sustainable farming that feeds everyone and remunerates food producers highly and (2) an export-oriented sector mainly dedicated to solidarity-based in-kind trade, while building the wherewithal to sustain prolonged economic attacks and internal sabotage affecting local food systems. At the very least, agroecological techniques and economic incentives help develop soil properties that can enhance subsequent farming and wider ecosystem conditions that create other healthful benefits for farming and wider communities. This, of course, is quite a tall order and only takes care of a small fraction of economic activities and environmental impacts, which would have to be coordinated with other activities and practices in similar ways. Such coordination could be facilitated by means of a socialist state or whatever other kinds of institutions make the most sense in given conjunctures and conditions. Regardless, these are some of the major strategy questions that can see the light of day by reclaiming state socialism and studying its people–environment histories with a dialectical and historical materialist eye.

DIALECTICAL AND HISTORICAL MATERIALISM AS AN OPEN-ENDED PROCESS OF MUTUAL TRANSFORMATIONS

The above is but one way of looking at the world of people–environment relations dialectically and through a historical materialist prism, instead of treating ecosystems and environments as static backgrounds or as unproblematically comparable. Environmental legacies are not mere background and do not just provide limits to or burdens on the present. They are affected by present impacts in refashioning the extent of their influence on the present condition. A dialectical understanding means paying attention to processes, interlinkages and, crucially, relations that are part of a whole as well as processes of mutual transformations (Harvey 1996, 46–68; Levins and Lewontin 1985).

To clarify further, dialectics as understood here is not based on identifying a synthesis of a thesis and its antithesis. Hegel, Marx and Engels held no such view, though many still think so even recently (e.g. Shimp 2009, repeating Heilbroner's mistakes). It has instead long been known that it was Fichte, building on Kant's ideas, who developed the synthesis formula (Mueller 1958). The widespread legend about Marxist dialectics can in part be blamed on some Marxists, certainly those involved in codifying what became the prevailing and state-sponsored version of dialectics in the USSR. That bears a decisive role in propagating such a formulaic, actually non-Marxian idea of dialectics. But many also continue to make a muddle of dialectics, as when they separate it from a relational view (that things can only be defined in relation to each other, like anarchism, communism, socialism and capitalism). Marx's, and also

Hegel's, version of dialectics is shot through with relationality already, so a term like 'relational dialectics' is wholly misinforming if pointing to Marxian dialectics. One can regard this as part of the deformation or degeneration of socialism in state power garb and leave the matter at that. Or we can take stock of the experiences, learn from them and struggle to ensure that dialectical materialism, just like the word communism, will no longer be distorted that way and will be turned to directly practical use in figuring out political strategies, resolving tensions, respecting and reconciling differences towards unity of purpose and action, and so on.

This may seem pedantic, but consider what kinds of policies can come out of a synthesis formula compared to a dialectical view. In the first, there are supposed to be two processes that contradict or conflict with each other and there is a resolution of that contradiction or conflict called a synthesis. So, one is supposed to look for things that are related but in contrast with each other. Applied to ecological processes or to how societies relate to environments, this becomes farcical if not dangerous. Is reforestation a synthesis of deforestation (thesis) and planting trees (antithesis)? What is the antithesis of a forest, a grassland, a desert, a wetland or some other kind of biome? Forests can also regenerate without planting trees and reforestation is not necessarily an ecological improvement, as one can have higher-biodiversity forests that are cut down, overtaken by cropland, and then, as a synthesis, replaced by low-biodiversity monocultural tree plantations as forests.

As conceived here, dialectical materialism is about looking out for and grasping how things relate to each other and change and about reckoning with mutual transformations that imply specifically human action and relationships between people and the components that comprise the rest of the universe. Reforestation, then, is a process of people relating differently to each other and at the same time to other beings within an ecosystem. This puts the emphasis on decision-making processes, on struggles over what sort of ecosystem is desirable, etc., rather than on finding out what is antithetical to what and forcing some sort of desired outcome (synthesis), other possibilities be damned.

Dialectical materialism is also a process of people transforming themselves in the process of transforming an ecosystem as part of a larger whole made up of interrelated societies and social interactions with and within ecosystems. There is no prescription, no foreclosed future, so whatever action is undertaken has no predictable outcomes or synthesis to another level. Dialectical materialism is an openness to outcomes being different from expectations, to maintaining the future as radically open with possibilities and to recognising how oneself is changed by the relationship with other beings or processes and with changes in the whole of which one is part. It is no recipe or formula. It is, rather, a set of methodological guidelines based on thinking about the world as always changing (processes, transformations), as things always defined in relation to other things and as part of sets of relations in a larger whole. This framework is a far more politically constructive alternative to the prevailing approaches of reductionism, monocausality, dualism, supremacism or objectivism, among other frameworks that fail to be socially critical and critically self-reflexive (Plumwood 1993).

A RESEARCH AGENDA TO DEVELOP QUESTIONS AND QUANDARIES FOR ECOSOCIALIST STRATEGIES

A historical and dialectical materialist framework, as outlined above, could be one way of proceeding, but with a critical reclaiming or interpretation of state socialism as intrinsically contradictory, ambivalent and transitory. This work is underlain by a three-pronged research agenda: (1) relational comparisons of state-socialist systems' environmental impacts that go beyond national-level analyses and refuse to treat such systems as monolithic or isolated; (2) multiple-scale studies of social causes of environmental degradation as well as environmental improvements within and beyond state-socialist countries (e.g. transboundary pollution, international preserves); (3) investigations into biophysical processes explaining environmental degradation or the attenuation of human impacts related to state-socialist policies and practices. Such a research agenda, informed by a critical reclaiming of state socialism and a historical and dialectical materialist framework, opens a way to formulate questions and highlight quandaries that must be faced when ecosocialist struggles gain momentum or succeed in redirecting society towards ecosocialism.

One way to enter this combined revisitation and preparatory activity is to learn from historical processes. There is no shortage of ideas about what can be learned, what the objectives should be, and what needs to be done differently for a future revolution or a world-systemic change towards socialism (e.g. Baer 2018; Bennholdt-Thomsen and Mies 1999; Kovel 2007). Yet too often it is underappreciated or even forgotten that the new revolutionary formation that became the USSR was invaded militarily soon after its establishment and that it was constantly besieged economically and threatened militarily by liberal democracies until its dissolution on 26 December 1991. Whatever strides were achieved socially and environmentally, and there were many, were thenceforth achieved under tremendous external pressures. At the same time, the Bolsheviks in power did not have the allegiance of most of the peasant majority. It is in this context of internal political weakness and enormous external pressures that the USSR's degeneration into an ever more centralised, paranoid, partially self-destructive and, for a time, terrorising set of institutions came about (Getty and Naumov 1999; Weiner 2017). In China the historical lesson was incorporated to develop strategies to ensure the backing of most peasants, but the revolutionaries had to contend with a long tradition of non-capitalistic commercial orientation as well (D'Mello 2009; Marks 1998).

These aspects alone should remind leftists of the high stakes and the immediate, murderous capitalist reprisals that accompany any successful or even potentially successful revolutionary activity. Out of the experience of the Russian Revolution, one could say that achieving socialism (and eventually communism) hinges at the very least on obtaining mass support and securing international coordination pre-empting deadly capitalist reaction. This requirement of at least neutralising the most militarily and economically powerful capitalist formations is something that Marx and Engels had pointed out by the 1880s (Rodney 2011). Subsequent revolutions nevertheless benefited greatly from the support of what survived of the USSR, even if eventually splits emerged, as with the PRC, Yugoslavia and Albania. From these splits there is much that still needs to be learned about the advantages and dangers of state power and political fragmentation in an international scene characterised by increasing centralisation and concentration of capitalist power embodied in gigantic corporations.

The development of technology and expansion of productive capacity were primary objectives in socialist states not only in raising the material well-being of the majorities whose lives were mainly awful in pre-revolutionary times, but also in terms of sheer survival under siege from 'the most advanced countries' (Weiner 1988, 233) and from inimical forces within state-socialist countries. A contradiction ensued from this general set of conditions that should be much more widely appreciated. The connection between developing the forces of production and living standards improvements and self-defence was made amply clear not too long after the first successful socialist revolution in what came and went as the USSR. Almost everything was subordinated in practice and ideologically to the defence of the only socialist country, conflating the USSR with world socialism, until the conceit was exposed most forcefully with the establishment of the PRC and the tensions erupted into conflict by the 1960s (not on account of Stalin, interestingly, but the anti-Stalinist Khrushchev faction).

The legitimate claim for self-defence and survival concealed internal strife and increasing centralisation of political decision-making processes, which were also precipitated by capitalist encirclement. All Bolshevik factions aimed for socialism to triumph and saw the USSR as threatened. The matter was especially urgent in the early decades, but the threat and massive pressures lasted even while chances of invasion were markedly reduced by means of the insanity of the nuclear warhead détente. One can point to the 1930s purges or even earlier, with the violent suppression of the Kronstadt Soviet (1921), as the beginning of the end or of the development of a degenerating system, but the state of siege was no mere Bolshevik hallucination or ideological expedient. All countries that underwent similar or similarly inspired revolutions faced similar challenges and attacks directly from liberal democracies or through their proxies. Constantly. The Cuban state, which as argued here is the only socialist state left, is a living example of what would happen to a successful revolution elsewhere if any of that calibre came about again.

The above comprise major social contradictions that are still rarely confronted by ecosocialists and other anti-capitalist leftists. They are manifested in three main ways. One is the unfavourable contexts wherein socialist movements were able to bring about revolutions and the socialist states (instead of, for example, confederations of workers councils) that emerged from those revolutions. The present neoliberal capitalist predominance throughout most of the world has exacerbated or newly caused global environmental degradation, and this constitutes a major set of conditions for current struggles. Another process is the socialism-undermining interconnectivity between state-socialist and capitalist countries, which are former or present colonial powers. This was a challenge in the past and will be so for any future successful revolution, including between socialist states or formations. A third major context is composed of pre-existing and newly formed internal social conflicts, including decolonisation and class struggles, which are always constituted through multiple forms of oppression. These are not going away and were sets of relations that posed great challenges to socialist movements, whether they succeeded or not in bringing about revolutions.

These three overarching processes combined in the past to undermine or redirect in unexpected negative directions whatever promising changes socialist revolutions or socialist states were undergoing. But they were also met with the resolve of many pushing against those destructive aspects and succeeding in eventually arriving at better living conditions and net positive environmental impacts. Nevertheless, those processes were and will be sources of contradictions for socialist projects.

One major contradiction to always keep in mind is building defences to fend off capitalist powers while bringing sufficient material well-being to all and less, not more, militarism. Part of this is also ensuring that basic infrastructure is functioning or even exists at all. The Norilsk smelters, the coal mines of the German Democratic Republic, the development of processing plants in Vietnam and the PRC's Great Leap Forward are among the numerous examples that followed from these pressures and represent attempts to improve living standards and the state of a country with available technologies.

214

The second main contradiction is in building the foundations of a classless society, inimical to endless capital accumulation, while having to exert the utmost effort to survive in a capitalist world economy. To start building such foundations, most socialist revolutions had to face and try to overcome the oppressive and at times ecologically unsustainable social structures in which many revolutionaries were raised. Many revolutions occurred in mainly agrarian or peasant-majority societies with semi-feudal arrangements heavily influenced by capitalist relations. The undermining and transformation of peasant systems had been well under way prior to those revolutions, and in most cases under the brutal and stultifying conditions of colonialism. Just to enable people to raise their material well-being required technological and economic resources unavailable in the countries where the revolutions succeeded. Having to import manufactures or technologically crucial parts often necessitated raw material export under often declining terms of trade. The raw material export aspect is one example of pressures to introduce or expand environmental damage. Something similar occurred by way of loan repayment and joint company schemes in the industrialised state-socialist countries, where environmentally unsound production was to a major degree to satisfy capitalist countries' market demands. Simply delinking from the capitalist world economy can lead to a drastic reduction in lifespans or worsening living standards, the stuff that creates the mass discontent useful in helping reactionaries restore capitalism.

These contradictory processes are much more relevant to and explanatory of environmental impact, whether positive or negative, rather than some problems intrinsic to state socialism. In other words, these are causal processes that can be made visible when letting go of liberal democratic presumptions or when resisting the temptation of judging what came out of socialist revolutions what socialism should be, instead of addressing the interrelationships of actually existing conditions, errors and atrocities, so as to be better prepared next time. In this light, the social and environmental improvements accomplished through socialist states ought to be even more appreciated and thoroughly studied.

What emerge from the application of the above framework are three broad, interrelated factors that shaped the ways in which the relationship between state socialism and the environment has evolved. One is the effects of the shifting context of socialist states in a capitalist world economy. Another is to be found in the changing interconnections between socialist states and capitalist powers, if not former colonial powers. A third is in the pre-existing and newly formed internal social conflicts faced in and between state-socialist countries following successful revolutions or systemic changes, including class struggles. These three processes inform overarching quandaries that need to be faced by any current socialist formation: building the foundations for a future classless and state-free society and the defences to fend off capitalist powers and survive within a capitalist world economy, while bringing sufficient material well-being to all and, in the process, less, not more, environmental harm. All this is necessary while at the same time ensuring close coordination and mutual support among socialist states and socialist movements. Addressing these quandaries involves two forms of interrelated but different struggles: a social and an ecological one. Socialist states exemplified this combined struggle. They still offer much not only in terms of signposts about what to prevent, but also of potentials to overcome capitalist relations in ecologically sustainable ways.

RECLAIMING STATE SOCIALISM

As Eduardo Galeano once put it, 'we are all invited to the world burial of socialism', but the 'funeral is for the wrong corpse' (Galeano 1991, 250). The struggle for socialism is just as crucial as ever, but the trick is how to combine anti-authoritarian socialist or socially egalitarian practices and theories with the need to build an ecologically minded and sustainable alternative to capitalism; and accomplish this in such a way as to avoid the pitfalls of state socialism and help build ecosocialist futures. More egalitarianism and environmental sustainability, even if not expressed as ecosocialism, is what millions of people seem to be demanding more and more even if this is diversely articulated, such as red-green alternatives, savordaya, ecological civilisation, democratic confederalism and the like. Socialist states could have a positive role in transitioning towards such biophysically and socially constructive alternatives. Those states, especially in Cuba, have demonstrated the potential to move in that direction.

Overall, the environmental record of state-socialist systems has been positive. In part this is due to the environmental sensibility pervading sections of what became ruling parties, particularly in the USSR. That sensibility endured even at the height of civil war and internal, mass-murderous repression. Socialist states like the PRC, which were turned into one-party governing institutions presiding over capitalist economies, have demonstrated an environmentally disastrous tendency. In the PRC the struggle within the party and society at large over the meaning and practice of ecological civilisation, which means returning to a path towards egalitarianism, seems decisive in stemming and reversing the environmentally destructive tide. To some extent this tendency towards greater environmental harm with more capitalist reform is evident when pro-capitalist technocrats within a ruling party have a greater say in running an economy within state-socialist countries, as happened in the USSR and the industrialised state-socialist countries in central and eastern Europe towards the end. However, the tendency was held in check by combinations of environmental movements, concerned or sympathetic high-ranking party officials and regional to local bureaucracies who stood to lose from environmental degradation. In countries like Cuba the situation was precipitated by a terrible economic downturn coupled with a throttling US embargo that is now 60 years old. But the environmentally constructive basis of present accomplishments is part of long-standing concerns with undoing the damaging and deadly environmental and social legacies of the capitalist past, including the struggle to overcome centuries of plantation economy practices grounded in sugar exports.

The lower ecological footprints in even the most industrialised forms of state socialism reflect a more constructive relationship with the biosphere. Biodiversity has been enhanced directly by extensive habitat and ecosystem protection measures as well as by successful afforestation and reforestation programmes, and, especially in the case of the USSR, ecological preserves. Soil protection and conservation have been effective and of lasting positive consequence, while organic farming techniques were starting to be increasingly favoured in the latter days of countries like the USSR. It is as if, with the dismantling of the USSR, the mantle has been taken up by Cuba. There the struggle for biophysically sustainable farming within the socialist state bore major successes with the development and spread of food production systems based on agroecological principles and traditional peasant knowledge systems. Thanks to these strides, Cuba is the epicentre of agroecological development and application. What is more, advances in urban farming are showing to the rest of the world how to integrate the urbanisation process with more environmentally sustainable living and direct grassroots participation. Socialist states also achieved lower ecological footprints by privileging mass transit and railways, and, in most instances, promoting recycling and minimising consumer waste. Air and water pollution in the industrialising state-socialist countries were addressed from the beginning, even if in fits and starts, owing to industrialising at breakneck speeds. Eventually, pollution abatement measures were increasingly more effective with time. At international levels, socialist states, especially the USSR, took the lead in long-range air pollution treaties, climatological research and climate change concerns, and desertification containment campaigns, among other environmental issues of global consequence. Lastly, socialist states can and have played crucial roles in ecological sensibilisation and mass environmental literacy by making education systems available to all, by means of diffusing scientific principles that are helpful towards understanding biophysical processes and by enabling biological conservation and naturalist movements and organisations to thrive and even gain influence over economic policy. Generally, it appears that the more socialist states give room to capitalist relations and to global capital flow, the greater the deleterious environmental impact.

There is much that can be built or rebuilt from the decades of experiences and successes in state-socialist countries. It is neither necessary nor advisable to start from scratch if the desire is to move as quickly as possible beyond capitalism. Rejecting state socialism is not much of an option, considering how state-socialist systems helped reduced or mitigate the destructive tendencies of capitalism and given the deterioration of social and ecological conditions worldwide, especially since the 1990s. What should be done instead is to revisit and learn from socialist states so as to build on their strengths and overcome their negative aspects. That is, each state-socialist case can be studied in context, as I have done here to some extent, to learn what worked well towards biophysical sustainability and determine the reasons for such successes. The same should be carried out regarding negative outcomes.

POTENTIAL ECOSOCIALIST PATHS AHEAD

No bloc centred on the USSR exists now, only a besieged Cuba and a few socialist governments, like Bolivia and Venezuela, under attack from liberal democracies and with little control over their national economies. The PRC has been in a counter-revolutionary phase since 1978 that necessitates support for the forces aiming to turn the tables, including within the CPC. Something similar has occurred or is occurring in Laos and Vietnam, and the rest of the socialist states have been wiped out since the early 1990s and replaced by capitalist client states. The role of ecosocialists and other like-minded people in the imperialist liberal democracies (mainly the NATO countries) and within the PRC is therefore especially important in overturning the balance of power globally, in supporting communists and genuine socialists in other countries and in allowing the flowering of ecosocialism where it has taken root most promisingly so far, as in Venezuela and Cuba. But it is just as important in contributing to the self-defence struggles of Indigenous peoples, whose lifeways are among the paragons for ecosocialism. In this respect, the late Edward Benton-Benai, co-founder of the American Indian Movement, in recounting Anishiinabeg Prophecy, put the matter in a way that is particularly apt for ecosocialists of any persuasion: 'the light skinned race will be given a choice between two roads. One road will be green and lush, and very inviting. The other road will be black and charred, and walking it will cut their feet' (Benton-Benai 1988, 93).

The Venezuelan case is one that brings to light these quandaries. There, ecosocialism has representation at the ministerial level (the first such in the world) and, as pointed out in Chapter 2, there are agroecology programmes and smallholder farmers' struggles for food sovereignty with the backing of state institutions; there are also promising projects featuring the establishment and coordination of communes and workers cooperatives with state institutions (Burbach and Piñeiro 2007; Ciccariello-Maher 2014). Instead of support for these projects and state initiatives, some fixate on the Bolivarian socialist government's political flaws and continued oil extraction (Bruno 2017; Machado 2017), in other words, for trying to survive on an exports-based oil economy that has been historically established and currently reinforced by capitalist forces within and abroad. While trying to improve living standards and relations with the environment, the Bolivarian socialist government is dealing with a national economy overwhelmingly in capitalist hands and with relentless military and economic attacks from the US and allied states. Criticisms of the Bolivarian socialists so far furnish no strategies or workable plans to deal with these quandaries, in contrast to some ecosocialists (Schwartzman and Saul 2015). The matter is not just about the survival of a socialist government. Promising life-affirming prospects for Afro-Venezuelans and Indigenous people, put in motion through the Bolivarian socialist process, are also at stake. The work of collectives like the Alberto Lovera Bolivarian Circle has been a shining illustration of critical support for the Bolivarian socialist project in countries plagued by governments intent on destroying it.

There are plenty of ideas and organising efforts, as indicated in Chapter 1, to learn from past state-socialist histories and improve upon them. For the most part they are based on rejecting state takeovers or institutional processes that subordinate worker control to central decision-making structures and insisting on worker and community self-organisation and self-management. Some resonate with what is advocated for here. That is, an appropriation of means of production (securing an economic base), an environmentally sustainable development of the forces of production that builds on existing subsistence perspectives (Bennholdt-Thomsen and Mies 1999) and the constant striving for economic democracy can be combined with identifying and availing oneself of possibilities emerging from constructive linkages to and intervention in existing state institutions (for examples, see Akuno and Nangwaya 2017; Kovel 2007, 263–75). This includes assemblies or similar processes where politically organised forces can bring learning experiences to communities struggling for self-determination. In some respects this harkens back to the Maoist mass line approach (D'Mello 2009) and could be useful in coordinating political forces within and outside the state, or a socialist state if one arises. It is unnecessary to reinvent the wheel with the great wealth of political organising histories and knowledge inherited from manifold socialist currents.

Investigation aiming to offer viable constructive actions are feasible if, instead of regarding state socialism as a form of degeneracy of some pious intent, one concentrates on the transitory, protean character of socialist government of the sort in Venezuela or Bolivia, where the objective is to build socialism. Some past state-socialist institutions, for instance, should be considered as a future reference, like the mass-mobilising scientific and naturalist organisations and people's control committees in the USSR, which were established shortly after the revolution. Such organising from below proved essential in the USSR in parrying technocrats and capitalism-tending reformists. Relative to Cuba currently, the same kind of methodology can be applied constructively if state socialism is understood as a contradictory phase that could go towards ecosocialism. In a state-socialist country, as in any other country, social struggles are not held in suspension. They continue, change and evolve as the balance of forces shifts and, biophysically, as impacts on the environment reshape the ways social relations connect to the rest of nature. In this reading, both material and political support for the socialist state's environmental and social advances in Cuba from countries like the US are crucial towards enabling ecosocialist prospects. The dire environmental situation in the PRC, as another example, can be overcome if the existing socialist forces within and outside the party prevail. The task of ecosocialist outsiders is to do the utmost to help those forces succeed (see Li 2016; 2017).

Another problem is engaging with political strategy and the necessary preparatory work for times when capitalist relations are on the wane. Part of this is a repeated failure to engage directly with technical expertise outside strictly social questions and to develop the means, in preparation to any systemic change, to build collectives of politically committed, egalitarian and technically capable specialists.

This is essential to confronting and resolving environmental challenges. Environmental destruction poses sets of complex dynamics, and the recent environmental reorientation among many radical leftist currents has been important in this respect. But radical leftist movements' predominant external criticism (if not at times rejection) of biophysical science frameworks and a near lack of systematically worked-out alternatives only facilitates the capitalist co-option of those sciences (see, for example, Gare 1993; Schwartzman 1996). Failure to rally technical scientists around radical causes or to become directly involved in producing the scientific knowledge and institutions that will be required for a transition to socialism is to court the sort of troubles the Bolsheviks faced right after the revolution. The Zapatista science initiative ConCiencias por la Humanidad is one example that addresses this challenge directly,[1] as well as the adoption of agroecological methods by the Landless Workers' Movement in Brazil (Borsatto and Souza-Esquerdo 2019). A similarly important development is the reconstitution of collectives like Science for the People[2] and Pandemic Research for the People,[3] both in the US.

Just as tough is overcoming current and historical divisions and antagonisms among leftists. These may be intractable because, as histories of socialism show, there are fundamental and possibly irreconcilable differences among socialist currents regarding objectives, irrespective of overlaps in ultimate ends regarding a classless society (which implies no state). To some extent the antagonism also stems from an age-old strife over centralist and bottom-up approaches. But this legitimate conundrum about political method often conceals a form of maximalist politics that leads to self-isolation. It is an inability to grasp the possibility that differences among radical leftists may be much smaller than between all radical leftists and the above-described politically reactionary camps. The matter is not as simple as presented here, certainly. There are persisting overlaps in practice, if not in ideology, between some leftist groups and reactionary formations. The issues of decolonisation, racialisation and masculinism are among these overlaps, when they are ignored or

---

1   https://conciencias.org.mx/.
2   https://scienceforthepeople.org/.
3   https://arerc.wordpress.com/pandemic-research-for-the-people/.

treated as secondary in importance to class differentiation, instead of seeing all of them as fundamentally intertwined with or the *sine qua non* of class struggle, as part of a differentiated unity.

Finally, since it hardly depends only on socialists, the even tougher challenge is to clarify to all directly involved as well as to external communities how projects that have little to no ostensibly socialist content can nevertheless be shown to promote socialist ends. Such can be among the lessons of state socialism. Branding such projects socialist, like cooperatives where decisions are made by external or internal party leaders, only plays into the hands of detractors. If the context is such that socialists must forgo some socialist principles to make gains towards a socialist future, this must be articulated explicitly. Debates must occur and be as free as possible in terms of demonstrating how contradictory practices cohere with the ultimate objective of bringing about socialism. But this may not be feasible, especially under military dictatorships, so other strategies that enable room for debate must be designed and thought of in advance, even if clandestinely and among few people, always with provisions made for restoring fully open discussions when circumstances allow and with the understanding that such discussions will not be effective if participants are not on equal terms. This is among the many lessons from the Bolshevik or, more recently, the Kurdistan Workers' Party experiences.

As stated in the Introduction, the process of state socialism turning into fully fledged socialism can be and has been upended to re-establish capitalism in one or another of its political variants, depending on the outcomes of internal and global dynamics. More positively, state socialism is a situation that can lead to a fully developed socialist society, with, among other processes, the socialisation of social reproduction and of the means of production (including resource access and production output), systematic redistribution of wealth to cover everyone's daily needs and workplace democracy prevailing, all leading to establishing communism, that is, state-free and egalitarian communities. That, not incidentally, does not entail the end of conflicts or tensions, nor of all forms of inequality. Nor does it naturally lead to the end of environmental destruction. A communist society must also develop the means to resolve social tensions in constructive ways and become ecologically sustainable, which is a complementary but different kind of struggle. It is not necessary to start from scratch, as many Indigenous peoples have developed constructive ways of addressing social tensions and destructive environmental impacts, certainly more constructive than capitalist or socialist state systems. However, such societies cannot be simply replicated, nor can their lifeways be superimposed on state-based systems. It is state-based systems, not just capitalist ones, that must be transformed and overcome, not only to arrive at much more constructive relations with the rest of nature, but also to stop the marginalisation and annihilation of state-free societies.

To repeat, addressing the above-identified contradictions involves social and environmental struggles. They are both multifaceted. The social struggle is to unify all egalitarianism-seeking forces, including within state institutions, to neutralise and overcome multiple forms of oppression and relations of domination. This is to bring about egalitarian, classless, state-free communities where they do not exist, and to build power with those already in existence. The ecological struggle is to impede environmental destruction, to foster life-promoting practices beyond those directly benefiting us and to find ways to live with compromised ecosystems that are irrevocably altered. These objectives imply unifying all forces, including within state institutions, seeking ecological sustainability and vying to avoid harming other species. The social and ecological struggles are not separate, but they are different. Bringing them together is an additional struggle and one that should not be taken for granted because one does not seamlessly flow into the other.

The current global environmental disasters cannot be confronted without social institutions capable of mobilising people and resources to that scale. Socialist states have that advantage and therefore remain relevant to the present struggles for an ecologically sustainable class- and state-free society. They are examples from which to learn and unlearn, through which to foresee and prepare and on which to build to overcome their flaws. Socialism, defined as the social ownership of the means of production with social equality, is a destination from which societies can evolve in a saner way than under capitalism. That evolution is not guaranteed to be towards an ecologically sustainable society. What previous forms of socialism demonstrated is how important relations are with the rest of nature and how such relations cannot be resolved by making changes only within society. Histories of socialist states demonstrate that both kinds of broad struggles must be waged at once (social and ecological) while at the same time confronting capitalist legacies of environmental devastation and internal (potentially violent) contradictions under a state of siege from outside forces.

If mobilisation through socialist states is done with environmentalist priorities, alongside social justice, much more can be done constructively and more rapidly. The crucial matter is keeping state institutions responsive, accountable from the bottom up. This is no easy task. Its feasibility is highly contingent on the constellation of forces within and outside each country, among other factors, like inherited economic structure and relative dependency on core capitalist economies, timing and conditions of state formation, etc. In capitalist countries this kind of mobilisation that can override profit-oriented interests is more difficult and historically remains to be demonstrated. On the other hand, the strides made in state-socialist countries have largely been unmade, to varying degrees, once socialist states were replaced by capitalist regimes. However, these are more contingent, historical trends, and they can only point to ways forward in relative terms. Such historical changes are important to take into consideration when devising ways to create a responsive, accountable set of state institutions, ideally to be converted into confederative bodies, not unlike what Rojava revolutionaries advocate for, as well as the EZLN and other similar movements. In Rojava and Chiapas there are examples of bottom-up decision-making processes about environmental conservation, promoting ecological understandings, devising ways to empower women (teaching jinealogy in Rojava, which is tied to ecological sustainability rather explicitly) and reconcile different cultural groups in ways that enable policy implementations including environmentally positive practices. These are in some ways spelled out in their very constitutions, even if practices may be less than stellar, largely because of highly inimical forces from without and within, particularly in the case of Rojava.

Writing a book about these issues is relatively easy, certainly compared to the much more arduous work of organising and taking direct action. May what is offered here at least contribute towards addressing the historical challenges of self-correction as a complement to struggles forthcoming, in the spirit of Amílcar Cabral's guiding principle that, 'We must constantly be more aware of the errors and mistakes we make so that we can correct our work and constantly do better in the service of our Party. The mistakes we make should not dishearten us, just as the victories we score should not make us forget our mistakes' (Cabral 1979 [1965], 226).

This is key to preventing even more catastrophes than have already happened and are happening. Capitalist societies, including liberal democracies, where mass destruction is largely reserved for those elsewhere, are the most destructive monstrosities in human history. To dismiss socialist states is in some ways to dismiss the prospects for a better world.

Postface

The Western Left, or at least a substantial part of it, hates all Left-wing countries and movements that have ever come to power ... It hates China and Vietnam, it hates Cuba and Venezuela. It wants to stay pure: doesn't want to govern, and it probably doesn't even want any Left wing country to exist ... When one governs, one makes errors, but how else to move forward? A big chunk of the Left in Western countries only wants to cry about how it is being marginalised, it hates taking risks. (Vltchek 2015, 184–5)

The assumption of responsibility for the language of the post-Soviet and the reclaiming of its vocabulary should start with an operation of negation; a refusal to use 'post Communist', 'post socialist', and 'post-Soviet' in the discourse of late capitalism. Instead, we should employ such terms as a 'time of interruption of socialism', 'time of regression', 'time of betrayal of progress', 'time of crude retro-capitalism', or simply 'the 1990s and 2000s' (Fiks 2007)

A corpse lies about, now displayed as a dismembered memory, now precariously stitched together. With the utmost contempt and tax-idermic care, the macabre spectacle is kept on display by its killers and by those feasting on its tattered and tarnished remains. For foe and fair-weather friend alike cherish remembrances and traces the corpse has left. Memories of its existence are terrifyingly useful, whether distorted or plainly invented to suit the justification of the day, if not for fear of its reincarnation and reconstitution from the menacing larvae that the assassins and scavengers constantly conjure up through their liberal dispensations of rights, property and war. But the assassins and scavengers ultimately behold manicured images of themselves, projected on what amounts to a lesser fraction of the many-flowered movements that continuously spring up, clamouring for a better life, if not a just future or just a chance for a life at all. It is my suspicion that the assassins and scavengers know this. Keeping the corpse alive therefore serves them all too well. Their purposefully misnamed and putrefied representation will linger as long as capitalist liberties and prerogatives prevail.

For the most part, those of a mindful constitution in the ecologically indebted worlds simply gawk at and acquiesce to the macabre spectacle. Sometimes they even abet its dismembering and reweaving. It would appear they are sometimes consciously complicit, but for the most part it is a matter of a debilitating lack of awareness and knowledge. There is anyway little solace and even less reward in becoming familiar with a reviled corpse unless one is prepared to join in the now customary collective revulsion. This is also expressed in the form of pre-emptive disgust for Bolshevism or Stalinism. The recitation is by now so thoroughly normalised and internalised that it is nearly compulsory, or at least felt as a reasonable expectation. This is all the more reason to lunge, without fear or remorse, into the histories and contexts hidden or refracted by the spectacle of the corpse. The rotting carcass must be exposed for what it really is, the tendency inhering capitalism to wreck what sustains us and to deny the satisfaction of needs and self-fulfilment potentialities of the world's majorities.

References

Abii, T., and P. Nwosu, P. 2009. 'The Effect of Oil-Spillage on the Soil of Eleme in Rivers State of the Niger-Delta Area of Nigeria'. *Research Journal of Environmental Sciences* 3 (3): 316–20.

Achard, Frédéric, Danilo Mollicone, Hans-Jürgen Stibig, Dmitry Aksenov, Lars Laestadius, Zengyuan Li, Peter Popatov and Alexey Yaroshenko. 2006. 'Areas of Rapid Forest-Cover Change in Boreal Eurasia'. *Forest Ecology and Management* 237: 322–34.

Agyeman, Julian, and Yelena Ogneva-Himmelberger, eds. 2009. *Environmental Justice and Sustainability in the Former Soviet Union.* Cambridge, MA: MIT Press.

Akuno, Kali, and Ajamu Nangwaya. 2017. 'Toward Economic Democracy, Labor Self-management and Self-determination', in Kali Akuno and Ajamu Nangwaya, with Cooperation Jackson (eds), *Jackson Rising: The Struggle for Economic Democracy and Black Self-determination in Jackson, Mississippi*, pp. 43–66. Cantley: Daraja Press.

Alix-Garcia, Jennifer, Catalina Munteanu, Na Zhao, Peter V. Potapov, Alexander V. Prishchepov, Volker C. Radeloff, Alexander Krylov and Eugenia Bragina. 2016. 'Drivers of Forest Cover Change in Eastern Europe and European Russia, 1985–2012'. *Land Use Policy* 59: 284–97.

Altieri, Miguel A., Nelso Companioni, Kristina Cañizares, Catherine Murphy, Peter Rosset, Martin Bourque and Clara I. Nicholls. 1999. 'The Greening of the "Barrios": Urban Agriculture for Food Security in Cuba'. *Agriculture and Human Values* 16: 131–40.

Altieri, Miguel A., and Fernando R. Funes-Monzone. 2012. 'The Paradox of Cuban Agriculture'. *Monthly Review* 63 (8): 23–33.

Altieri, Miguel A., Fernando R. Funes-Monzone and Paulo Petersen. 2012. 'Agroecologically Efficient Agricultural Systems for Smallholder Farmers: Contributions to Food Sovereignty'. *Agronomy for Sustainable Development* 32: 1–13.

Álvarez, A., J. Estévez Alvarez, C. Nascimento, I. González, O. Rizo, L. Carzola, R. Ayllón Torres and J. Pascual. 2017. 'Lead Isotope Ratios in Lichen Samples Evaluated by ICP-ToF-MS to Assess Possible Atmospheric Pollution Sources in Havana, Cuba'. *Environmental Monitoring & Assessment* 189 (1): 1–8.

Amin, Samir. 2018. *Modern Imperialism, Monopoly Finance Capital, and Marx's Law of Value.* New York: Monthly Review Press.

Anderson, Kevin B. 2010. *Marx at the Margins: On Nationalism, Ethnicity, and Non-Western Societies.* Chicago: University of Chicago Press.

Angus, Ian. 2016. *Facing the Anthropocene: Fossil Capitalism and the Crisis of the Earth System.* New York: Monthly Review Press.

Arrighi, Giovanni, Terence K. Hopkins and Immanuel Wallerstein. 1989. *Anti-systemic Movements*. London: Verso.

Baer, Hans. 2018. *Democratic Eco-Socialism as a Real Utopia: Transitioning to an Alternative World System*. New York: Berghahn.

Baracca, Angelo, and Rosella Franconi. 2016. *Subalternity vs. Hegemony: Cuba's Outstanding Achievements in Science and Biotechnology, 1959–2014*. Basel: Springer.

Barr, Brenton M. 1988. 'Perspectives on Deforestation in the USSR', in John F. Richards and Richard P. Tucker (eds), *World Deforestation in the Twentieth Century*, pp. 230–61. Durham, NC: Duke University Press.

Bedford, D., and D. Irving-Stephens. 2000. *The Tragedy of Progress: Marxism, Modernity, and the Aboriginal Question*. Halifax, NS: Fernwood.

Beik, Doris, and Paul Beik. 1993. *Flora Tristan, Utopian Feminist: Her Travel Diaries and Personal Crusade*. Bloomington: Indiana University Press.

Bekenov, A.B., I.A. Grachev and E.J. Milner-Gulland. 1998. 'The Ecology and Management of the Saiga Antelope in Kazakhstan'. *Mammal Review* 28: 1–52.

Bennholdt-Thomsen, Veronika, and Maria Mies. 1999. *The Subsistence Perspective: Beyond the Globalised Economy*. London: Zed Books.

Benton-Benai, Edward. 1988. *The Mishomis Book: The Voice of the Ojibway*. Hayward: Indian Country Communications.

Bentzen, Jeanet, Nicolai Kaarsen and Asger Moll Wingender. 2013. 'The Timing of Industrialization across Countries', University of Copenhagen Department of Economics Discussion Paper No. 13–17, https://ssrn.com/abstract=2391283 (accessed 28 November 2020).

Benz, Andreas. 2020. 'The Greening of the Revolution: Changing State Views on Nature and Development in Cuba's Transforming Socialism'. *GAIA: Ecological Perspectives for Science & Society* 29 (4): 243–8.

Berend, Iván T. 1996. *Central and Eastern Europe, 1944–1993: Detour from the Periphery to the Periphery*. Cambridge: Cambridge University Press.

Betancourt, Mauricio. 2020. 'The Effect of Cuban Agroecology in Mitigating the Metabolic Rift: A Quantitative Approach to Latin American Food Production'. *Global Environmental Change* 63: 102075.

Bin, Shui, and Robert C. Harriss. 2006. 'The Role of CO_2 Embodiment in US–China Trade'. *Energy Policy* 34 (8): 4063–8.

Biney, Ama. 2018. 'Madmen, Thomas Sankara and Decoloniality in Africa', in Amber Murray (ed.), *A Certain Amount of Madness: The Life, Politics and Legacies of Thomas Sankara*, pp. 127–46. London: Pluto Press.

Birchall, Ian. 1997. *Spectre of Babeuf*. Houndmills: Palgrave Macmillan.

Birman, Igor. 1989. *Personal Consumption in the USSR and the USA*. New York: Palgrave Macmillan.

Birmili, W., K. Schepanski, A. Ansmann, G. Spindler, I. Tegen, B. Wehner, A. Nowak, E. Reimer, I. Mattis, K. Müller, E. Brüggemann, T. Gnauk, H. Herrmann, A. Wiedensohler, D. Althausen, A. Schladitz, T. Tuch and G. Löschau. 2008. 'A Case of Extreme Particulate Matter Concentrations over Central Europe Caused by

References

Dust Emitted over the Southern Ukraine'. *Atmospheric Chemistry and Physics* 8: 997–1016.

Biró, Marianna, Bálint Czúcz, Ferenc Horváth, András Révész, Bálint Csatári and Zsolt Molnár. 2013. 'Drivers of Grassland Loss in Hungary during the Post-socialist Transformation (1987–1999)'. *Landscape Ecology* 28: 789–803.

Blackburn, Robin. 1991. 'Fin de Siècle: Socialism after the Crash', in Robin Blackburn (ed.), *After the Fall: The Failure of Communism and the Future of Socialism*, pp. 173–249. New York: Verso.

Blackburn, Robin. 2000. 'Putting the Hammer down on Cuba'. *New Left Review* 2000 (4): 5–36.

Blackburn, Robin. 2006. 'Haiti, Slavery, and the Age of the Democratic Revolution'. *The William and Mary Quarterly* 63: 643–74.

Blaikie, Piers. 1985. *The Political Economy of Soil Erosion in Developing Countries*. Essex: Longman.

Boden, Thomas A., and Robert J. Andres. 2014. 'Fossil-Fuel CO_2 Emissions by Nation'. Oak Ridge: Carbon Dioxide Information Analysis Center, National Laboratory, US Department of Energy. https://cdiac.ess-dive.lbl.gov/trends/emis/tre_coun.html (accessed 6 November 2020).

Boden, Thomas A., G. Marland and Robert J. Andres. 2011. *Global, Regional, and National Fossil-Fuel CO2 Emissions*. Oak Ridge: Carbon Dioxide Information Analysis Center, National Laboratory, US Department of Energy.

Boland, Rosita. 2017. 'Death from Below in the World's Most Bombed Country'. *The Irish Times*, 13 May 2017. www.irishtimes.com/news/world/asia-pacific/death-from-below-in-the-world-s-most-bombed-country-1.3078351 (accessed 18 May 2020).

Bolt, Jutta, Robert Inklaar, Herman de Jong and Jan Luiten van Zanden. 2018. 'Maddison Project Database, version 2018'. www.rug.nl/ggdc/historicaldevelopment/maddison/releases/maddison-project-database-2018?lang=en (accessed 27 January 2021).

Böröcz, József. 1992. 'Dual Dependency and Property Vacuum: Social Change on the State Socialist Semiperiphery'. *Theory and Society* 21: 77–104.

Borsatto, Ricardo Serra, and Vanilde F. Souza-Esquerdo. 2019. 'MST's Experience in Leveraging Agroecology in Rural Settlements: Lessons, Achievements, and Challenges'. *Agroecology and Sustainable Food Systems* 43 (7–8): 915–35.

Bot, Alexandra J., Freddy O. Nachtergaele and Anthony Young. 2000. *Land Resource Potential and Constraints at Regional and Country Levels*. World Soil Resources Report 90. Rome: Food and Agriculture Organization of the United Nations. www.fao.org/3/a-x7126e.pdf (accessed 11 November 2020).

Bragina, Eugenia V., Volker C. Radeloff, Matthias Baumann, Kelly Wendland, Tobias Kuemmerle and Anna M. Pidgeon. 2015a. 'Effectiveness of Protected Areas in the Western Caucasus before and after the Transition to Post-socialism'. *Biological Conservation* 184: 456–64.

Bragina, Eugenia V., A.R. Ives, Anna M. Pidgeon, Tobias Kuemmerle, Leonid M. Baskin, Y.P. Gubar, Maria Piquer-Rodríguez, N.S. Keuler, Varos G. Petrosyan and Volker C. Radeloff. 2015b. 'Rapid Declines of Large Mammal Populations after the Collapse of the Soviet Union'. *Conservation Biology* 29 (3): 844–53.

Bragina, Eugenia V., Anthony R. Ives, Anna M. Pidgeon, Linas Balčiauskas, Sándor Csányi, Pavlo Khoyetskyy, Katarina Kysucká, Juraj Lieskovsky, Janis Ozolins, Tiit Randveer, Přemysl Štych, Anatoliy Volokh, Chavdar Zhelev, Elzbieta Ziółkowska and Volker C. Radeloff. 2018. 'Wildlife Population Changes across Eastern Europe after the Collapse of Socialism'. *Frontiers in Ecology and the Environment* 6 (2): 77– 81.

Brain, Stephen. 2011. *Song of the Forest: Russian Forestry and Stalinist Environmentalism, 1905–1953*. Pittsburgh: University of Pittsburgh Press.

Brain, Stephen. 2016. 'The Appeal of Appearing Green: Soviet-American Ideological Competition and Cold War Environmental Diplomacy'. *Cold War History* 16 (4): 443–62.

Braunthal, Julius. 1967a. *History of the International – Volume 1: 1864–1914*. New York: Praeger.

Braunthal, Julius. 1967b. *History of the International – Volume 2: 1914–1943*. New York: Praeger.

Breyfogle, Nicholas B. 2015. 'At the Watershed: 1958 and the Beginnings of Lake Baikal Environmentalism'. *The Slavonic and East European Review* 93 (1): 147–80.

Brooks, Elizabeth, and Jody Emel. 1995. 'The Llano Estacado of the American Southern High Plains', in Jeanne X. Kasperson, Roger E. Kasperson and Billie L. Turner II (eds), *Regions at Risk: Comparisons of Threatened Environments*, pp. 255–303. Tokyo: United Nations University Press.

Broughton, Edward. 2005. 'The Bhopal Disaster and Its Aftermath: A Review'. *Environmental Health: A Global Access Science Source* 4 (1): 6.

Brown, Kate. 2013. *Plutopia: Nuclear Families, Atomic Cities, and the Great Soviet and American Plutonium Disasters*. Oxford: Oxford University Press.

Brown, Lester R., and Edward C. Wolf. 1984. *Soil Erosion: Quiet Crisis in the World Economy*. Washington, DC: Worldwatch Institute.

Bruno, Andy. 2016. *The Nature of Soviet Power. An Arctic Environmental History*. Cambridge: Cambridge University Press.

Bruno, Andy. 2017. 'Can Socialism Save the Planet?' *EdgeEffects*. https://edgeeffects. net/socialism-environment/ (accessed 19 June 2020).

Bruno, Andy. 2018. 'How a Rock Remade the Soviet North: Nepehline in the Khibiny Mountains', in Nicholas Breyfogle (ed.), *Eurasian Environments: Nature and Ecology in Eurasian History*, pp. 147–54. Pittsburgh: University of Pittsburgh Press.

Burbach, Roger, and Camila Piñeiro. 2007. 'Venezuela's Participatory Socialism'. *Socialism and Democracy* 21 (3): 181–200.

Buzmakov, Sergei, Darya Egorova and Evgeniia Gatina. 2019. 'Effects of Crude Oil Contamination on Soils of the Ural Region'. *Journal of Soils Sediments* 19: 38–48.

References

Cabello, Juan José, Dunia Garcia, Alexis Sagastume, Rosario Priego, Luc Hens and Carlo Vandecasteele. 2012. 'An Approach to Sustainable Development: The Case of Cuba'. *Environment Development and Sustainability* 14: 573–91.

Cabral, Amílcar. 1979 [1965]. *Unity and Struggle: Speeches and Writings*. Texts Selected by the PAIGC. Translated by Michael Wolfers. New York: Monthly Review Press.

Cameron, Sarah. 2018. *The Hungry Steppe: Famine, Violence, and the Making of Soviet Kazakhstan*. Ithaca, NY: Cornell University Press.

Cattaneo, Claudio, and Salvatore Engel-Di Mauro. 2015. 'Urban Squats as Eco-Social Resistance to and Resilience in the Face of Capitalist Relations: Case Studies from Barcelona and Rome'. *Partecipazione e Conflitto* 8 (2): 343–66.

Ceballos, Gerardo, Paul R. Ehrlich and Rodolfo Dirzo. 2017. 'Population Losses and the Sixth Mass Extinction'. *Proceedings of the National Academy of Sciences* 114 (30): E6089–E6096.

Ceddia, M. Graziano. 2020. 'The Super-Rich and Cropland Expansion via Direct Investments in Agriculture'. *Nature Sustainability* 3: 312–18.

Cereseto, Shirley, and Howard Waitzkin. 1986. 'Capitalism, Socialism, and the Physical Quality of Life'. *International Journal of Health Services* 16 (4): 643–58.

César, Filipa. 2018. 'Meteorisations. Reading Amílcar Cabral's Agronomy of Liberation'. *Third Text* 32 (2–3): 254–72.

Chase-Dunn, Christopher. 1982. 'Socialist States in the Capitalist World Economy'. In Christopher Chase-Dunn (ed.), *Socialist States in the World-System*, pp. 21–55. Beverly Hills: SAGE.

Cheesman, Oliver D. 2004. *Environmental Impacts of Sugar Production: The Cultivation and Processing of Sugarcane and Sugar Beet*. Wallingford: CABI.

Chen Hongyan, Xuyin Yuan, Tianyuan Li, Sun Hu, Junfeng Ji and Cheng Wang. 2016. 'Characteristics of Heavy Metal Transfer and Their Influencing Factors in Different Soil-Crop Systems of the Industrialization Region, China'. *Ecotoxicology and Environmental Safety* 126: 193–201.

Chen Ying, Jiahua Pan and Laihui Xie. 2011. 'Energy Embodied in Goods in International Trade of China: Calculation and Policy Implications'. *Chinese Journal of Population Resources and Environment* 9 (1): 16–32.

Chendev, Yuri G., Thomas. J. Sauer, Guillermo H. Ramirez and Charles L. Burras. 2015. 'History of East European Chernozem Soil Degradation: Protection and Restoration by Tree Windbreaks in the Russian Steppe'. *Sustainability* 7: 705–24.

Chernogaeva, G.M., V. Ginzburg, S. Paramonov, B. Pastukhov and O. Lysak. 2009. 'Integrated Background Monitoring of Environmental Pollution in Russia'. *Russian Meteorology and Hydrology* 34: 301–7.

Chomsky, Noam, and Robert Pollin. 2020. *Climate Crisis and the Global Green New Deal*. London: Verso.

Chu, Pei-Yi. 2018. 'Encounters with Permafrost: The Rhetoric of Conquest and Processes of Adaptation in the Soviet Union', in Nicholas Breyfogle (ed.), *Eurasian*

Environments: Nature and Ecology in Eurasian History, pp. 165–84. Pittsburgh: University of Pittsburgh Press.

CIA (Central Intelligence Agency, US). 1976. 'Amount and Percentage of World Population Dominated by the Soviet Union or under Communist Regimes'. www.cia. gov/library/readingroom/docs/DOC_0000969777.pdf (accessed 7 November 2020).

Ciccariello-Maher, George. 2014. 'Building the Commune: Insurgent Government, Communal State'. *South Atlantic Quarterly* 113 (4): 791–806.

Coase, Ronald, and Ning Wang. 2013. 'How China Became Capitalist'. *Cato Policy Report* 35 (1): 7–10. www.cato.org/sites/cato.org/files/serials/files/policy-report/ 2013/1/cprv35n1–1.pdf (accessed 7 October 2020).

Cockshott, Paul W., and Allin Cottrell. 1993. *Towards a New Socialism*. Nottingham: Spokesman, Bertrand Russell House.

Cole, George D.H. 1953. *A History of Socialist Thought: Volume I – The Forerunners 1789–1850*. London: MacMillan.

Cole, George D.H. 1960. *The Second International 1889–1914, Part II*. London: MacMillan.

Cole, George D.H. 1961. *Marxism and Anarchism 1850–1890*. London: Macmillan.

Colwell, Mark A., Alexander V. Dubynin, Andrei Yu. Koroliuk and Nikolai A. Sobolev. 1997. 'Russian Nature Reserves and Conservation of Biological Diversity'. *Natural Areas Journal* 17 (1): 56–68.

Coronel Vargas, Gabriela, William W. Au and Alberto Izzotti. 2020. 'Public Health Issues from Crude-Oil Production in the Ecuadorian Amazon Territories'. *Science of The Total Environment* 719: 134647. https://doi.org/10.1016/j. scitotenv.2019.134647 (accessed 4 November 2020).

Crawford, Neda. 2019. 'Pentagon Fuel Use, Climate Change, and the Costs of War'. https://watson.brown.edu/costsofwar/files/cow/imce/papers/2019/Pentagon%20 Fuel%20Use,%20Climate%20Change%20and%20the%20Costs%20of%20War%20 Final.pdf (accessed 21 January 2021).

Crippa, Monica, Gabriel Oreggioni, Diego Guizzardi, Marilena Muntean, Edwin Schaaf, Eleonora Lo Vullo, Efibio Solazzo, Fabio Monforti-Ferrario, Jos G.J. Olivier and Elisabetta Vignati. 2019. *Fossil CO$_2$ and GHG Emissions of All World Countries: 2019 Report, EUR 29849 EN*. Luxembourg: Publications Office of the European Environmental Agency.

Cvitanović, Marin, George Alan Blackburn and Martin Rudbeck Jepsen. 2016. 'Characteristics and Drivers of Forest Cover Change in the Post-socialist Era in Croatia: Evidence from a Mixed-Methods Approach'. *Regional Environmental Change* 16: 1751–63.

Cvitanović, Marin, Ivana Lučev, Borna Fürst-Bjeliš, Lana Slavuj Borčić, Suzana Horvat and Luka Valožić. 2017. 'Analyzing Post-socialist Grassland Conversion in a Traditional Agricultural Landscape: Case Study Croatia'. *Journal of Rural Studies* 51: 53–63.

References

D'Mello, Bernard. 2009. 'What Is Maoism?' *Economic and Political Weekly* 44 (47): 37–48.

David-West, Alzo. 2011. 'Between Confucianism and Marxism-Leninism: Juche and the Case of Chŏng Tasan'. *Korean Studies* 35: 93–121.

Davis, Steven J., and Ken Caldeira. 2010. 'Consumption-Based Accounting of CO_2 Emissions'. *Proceedings of the National Academy of Sciences* 107 (12): 5687–92.

Davis, Thomas. 2000. *Sustaining the Forest the People and the Spirit*. New York: State University of New York Press.

DeBardeleben, Joan. 1992. 'The New Politics in the USSR: The Case of the Environment', in John M. Stewart (ed.), *Soviet Environment: Problems, Politics, Prospects*, pp. 64–87. Cambridge: Cambridge University Press.

DEFRA. 2017. *Food Statistics in Your Pocket 2017: Global and UK Supply*. London: UK Department for Environment, Food and Rural Affairs.

Democratic Socialist Party. 1999. *Environment, Capitalism and Socialism*. Sydney: Resistance Books.

Dhara, V. Ramana, and Rosaline Dhara. 2002. 'The Union Carbide Disaster in Bhopal: A Review of the Health Effects'. *Archives of Environmental Health* 57 (5): 391–404.

Dickens, Peter. 1992. *Society and Nature: Towards a Green Social Theory*. Philadelphia: Temple University Press.

Dietrich, Matthew Amy Wolfe, Michelle Burke and Mark P.S. Krekeler. 2019. 'The First Pollution Investigation of Road Sediment in Gary, Indiana: Anthropogenic Metals and Possible Health Implications for a Socioeconomically Disadvantaged Area'. *Environment International* 128: 175–92.

Dietz, Thomas, and Eugene A. Rosa. 1994. 'Rethinking the Environmental Impacts of Population, Affluence and Technology'. *Human Ecology Review* 1 (2): 277–300.

Dirlik, Arif. 1991. *Anarchism in the Chinese Revolution*. Berkeley: University of California Press.

Dominick, Raymond. 1998. 'Capitalism, Communism, and Environmental Protection: Lessons from the German Experience'. *Environmental History* 3 (3): 311–32.

Du Bois, William E.B. 1945. *Color and Democracy: Colonies and Peace*, in Henry L. Gates (ed.), *The World and Africa and Color and Democracy: The Oxford W.E.B. Du Bois*. Oxford: Oxford University Press.

Dudenkov, Staly V. 1985. 'The Recycling of the Wastes of Production and Consumption as an Aspect of the Environmental Protection in the USSR', in Kriton Curi (ed.), *Appropriate Waste Management for Developing Countries*, pp. 95–100. Boston: Springer.

Dudiak, Nataliia Vasylivna, Vitalii Ivanovich Pichura, Larisa Aleksandrovna Potravka and Alexander Alexandrovich Stroganov. 2020. 'Spatial Modeling of the Effects of Deflation Destruction of the Steppe Soils of Ukraine'. *Journal of Ecological Engineering* 21 (2): 166–77.

Dung, Elisha, Leonard Bombom and Tano Agusomu. 2008. 'The Effects of Gas Flaring on Crops in the Niger Delta, Nigeria'. *GeoJournal* 73 (4): 297–305.

235

EC-JRC (European Commission, Joint Research Centre). 2019. 'Emissions Database for Global Atmospheric Research (EDGAR), release EDGAR v5.0 (1970–2015)'. https://edgar.jrc.ec.europa.eu/overview.php?v=50_GHG (accessed 7 November 2020).

EC-JRC (European Commission, Joint Research Centre). 2020. 'Emissions Database for Global Atmospheric Research (EDGAR), release EDGAR v5.0 (1970–2015)'. https://edgar.jrc.ec.europa.eu/overview.php?v=50_AP (accessed 7 November 2020).

Eckhardt, Wolfgang. 2016. *The First Socialist Schism: Bakunin vs. Marx in the International Working Men's Association*. Translated by Robert M. Homsi, Jesse Cohn, Cian Lawless, Nestor McNab and B. Moreel. Oakland: P.M. Press.

Eckholm, Erik. 1976. *Losing Ground: Environmental Stress and World Food Prospects*. New York: W.W. Norton.

Edelstein, Michael R. 2012. 'Death and Rebirth Island: Secrets in the U.S.S.R.'s Culture of Contamination', in Michael R. Edelstein, Astrid Cerny and Abror Gadaev (eds), *Disaster by Design: The Aral Sea and Its Lessons for Sustainability*, pp. 37–52. Bingley: Emerald Group.

Eisenman, Joshua. 2018. *Red China's Green Revolution: Technological Innovation, Institutional Change, and Economic Development Under the Commune*. New York: Columbia University Press.

Elie, Marc. 2015. 'Formulating the Global Environment: Soviet Soil Scientists and the International Desertification Discussion, 1968–91'. *The Slavonic and East European Review* 93 (1): 181–204.

Elie, Marc. 2018. 'Desiccated Steppes: Drought and Climate Change in the USSR, 1960s–1980s', in Nicholas Breyfogle (ed.), *Eurasian Environments: Nature and Ecology in Eurasian History*, pp. 75–93. Pittsburgh: University of Pittsburgh Press.

Elie, Marc, and Laurent Coumel. 2013. 'A Belated and Tragic Ecological Revolution: Nature, Disasters, and Green Activists in the Soviet Union and the Post-Soviet States, 1960s–2010s'. *The Soviet and Post-Soviet Review* 40: 157–65.

Engel-Di Mauro, Salvatore. 2002. 'Gender Relations, Political Economy, and the Ecological Consequences of State-Socialist Soil Science'. *Capitalism Nature Socialism* 13 (3): 92–117.

Engel-Di Mauro, Salvatore. 2003. 'Disaggregating Local Knowledge: The Effects of Gendered Farming Practices on Soil Fertility and Soil Reaction in SW Hungary'. *Geoderma* 111 (3–4): 503–20.

Engel-Di Mauro, Salvatore. 2006. 'From Organism to Commodity: Gender, Class, and the Development of Soil Science in Hungary, 1900–1989'. *Environment and Planning D: Society and Space* 24: 215–29.

Engel-Di Mauro, Salvatore. 2014. *Ecology, Soils, and the Left: An Eco-Social Approach*. New York: Palgrave Macmillan.

Engel-Di Mauro, Salvatore. 2018. 'Soils in Eco-Social Context: Soil pH and Social Relations of Power in a Northern Drava River Floodplain Agricultural Area', in

References

Rebecca Lave, Christine Biermann and Stuart Lane (eds), *The Palgrave Handbook of Critical Physical Geography*, pp. 393–419. New York: Palgrave.

Fateev, V.N., S.A. Grigoriev and E.A. Seregina. 2020. 'Hydrogen Energy in Russia and the USSR'. *Nanotechnology Russia* 15: 256–72.

Febles-González, J., M. Vega-Carreño, A. Tolón-Becerra and X. Lastra-Bravo. 2012. 'Assessment of Soil Erosion in Karst Regions of Havana, Cuba'. *Land Degradation & Development* 23 (5): 465–74.

Fernandez, Margarita, Justine Williams, Galia Figueroa, Garrett Graddy-Lovelace, Mario Machado, Luis Vazquez, Nilda Perez, Leidy Casimiro, Graciela Romero and Fernando Funes-Aguilar. 2018. 'New Opportunities, New Challenges: Harnessing Cuba's Advances in Agroecology and Sustainable Agriculture in the Context of Changing Relations with the United States'. *Elementa Science of the Anthropocene* 6: 76.

Feshbach, Murray, and Alfred Friendly Jr. 1993. *Ecocide in the USSR: Health and Nature under Siege*. New York: Basic Books.

Figge, Lukas, Kay Oebels and Astrid Offermans. 2017. 'The Effects of Globalization on Ecological Footprints: An Empirical Analysis'. *Environment, Development and Sustainability* 19: 863–76.

Fiks, Yevgeniy. 2007. 'Responsibilities of the Post-Soviet Artist'. *Moscow Art Magazine*. http://moscowartmagazine.com/issue/42/article/843 (accessed 17 January 2021).

Finkelstein, Norman G. 2003. *Image and Reality of the Israel–Palestine Conflict*. New York: Verso Books.

Fitz, Don. 2020. *Cuban Healthcare: The Ongoing Revolution*. New York: Monthly Review Press.

Foster, John Bellamy. 2000. *Marx's Ecology*. New York: Monthly Review Press.

Foster, John Bellamy. 2015. 'Late Soviet Ecology and the Planetary Crisis'. *Monthly Review* 67 (2): 1–20.

Frank, André Gunder. 1977. 'Long Live Transideological Enterprise! The Socialist Economies in the Capitalist International Division of Labor'. *Review (Fernand Braudel Center)* 1 (1): 91–140.

Frank, André Gunder. 1989. 'The Socialist Countries in the World Economy: The East–South Dimension', in Brigitte H. Schultz and William W. Hansen (eds), *The Socialist Bloc and the Third World: The Political Economy of East–South Relations*, pp. 9–26. London: Westview Press.

French, C., M. Becker and B. Lindsay. 2010. 'Havana's Changing Urban Agriculture Landscape: A Shift to the Right?' *Journal of Agriculture, Food Systems, and Community Development* 1 (2): 155–65.

Frey, R. Scott. 2012. 'The E-waste Stream in the World-System'. *Journal of Globalization Studies* 3 (1): 79–94.

Gadaev, Abror, and Zikrilla Yasakov. 2012. 'An Overview of the Aral Sea Disaster', in Michael R. Edelstein, Astrid Cerny and Abror Gadaev (eds), *Disaster by Design: The Aral Sea and Its Lessons for Sustainability*, pp. 5–16. Bingley: Emerald Group.

Galeano, E. 1991. 'A Child Lost in the Storm', in Robin Blackburn (ed.), *After the Fall: The Failure of Communism and the Future of Socialism*, pp. 250–4. New York: Verso.

Gare, Arran. 1993. 'Soviet Environmentalism: The Path Not Taken'. *Capitalism Nature Socialism* 4 (4): 69–88.

Gare, Arran. 2002. 'The Environmental Record of the Soviet Union'. *Capitalism Nature Socialism* 13 (3): 52–72.

Gare, Arran. 2012. 'China and the Struggle for Ecological Civilization'. *Capitalism Nature Socialism* 23 (4): 10–26.

Gare, Arran. 2016. *Philosophical Foundations of Ecological Civilization: A Manifesto for the Future.* New York: Routledge.

Gare, Arran. 2020. 'The Eco-Socialist Roots of Ecological Civilisation'. *Capitalism Nature Socialism.* https://doi.org/10.1080/10455752.2020.1751223.

Gareau, Brian J. 2013. *From Precaution to Profit: Contemporary Challenges to Environmental Protection in the Montreal Protocol.* New Haven: Yale University Press.

Getty, J. Arch, and Oleg Naumov. 1999. *The Road to Terror: Stalin and the Self-destruction of the Bolsheviks, 1932–1939.* New Haven: Yale University Press.

Gille, Zsuzsa. 2004. 'Europeanising Hungarian Waste Policies: Progress or Regression?' *Environmental Politics* 13 (1): 114–34.

Gills, Barry. 1992. 'North Korea and the Crisis of Socialism: The Historical Ironies of National Division'. *Third World Quarterly* 13 (1): 107–30.

Glantz, Mickey. 2007. 'Aral Sea Basin: A Sea Dies, a Sea also Rises'. *Ambio* 36: 323–7.

Global Footprint Network. 2019. 'Country Trends'. https://api.footprintnetwork.org/v1/data/5001/all/BCtot,EFCtot (accessed 8 November 2020).

Global Witness. 2020. 'Defending Tomorrow: The Climate Crisis and Threats against Land and Environmental Defenders'. www.globalwitness.org/documents/19938/Defending_ Tomorrow_EN_high_res_-_July_2020.pdf (accessed 26 October 2020).

Goldman, Marshall I. 1972. *The Spoils of Progress: Environmental Pollution in the Soviet Union.* Boston: MIT Press.

Goldman, Marshall I. 1988. 'The Development of Environmental Policies in the United States and the Soviet Union'. *Pace Environmental Law Review* 5 (2): 455–62.

Goldman, Marshall I. 1992. 'Environmentalism and Nationalism: An Unlikely Twist in an Unlikely Direction', in John M. Stewart (ed.), *The Soviet Environment: Problems, Policies, and Politics*, pp. 1–10. Cambridge: Cambridge University Press.

Golosov, Valentin, and Vladimir Belyaev. 2013. 'The History and Assessment of Effectiveness of Soil Erosion Control Measures Deployed in Russia'. *International Soil and Water Conservation Research* 1 (2): 26–35.

Gómez, Tania M. 2019. 'Logros y Desafíos de la Red Universitaria REDMA: Por la Sostenibilidad Ambiental de la Educación Superior en Cuba'. [Achievements and Challenges of REDMA University Network: For Cuban Higher Education's Environmental Sustainability.] *Revista Luna Azul* 49: 109–25.

References

González Novo, M., and A. Castellanos Quintero. 2014. 'Havana', in Graeme Thomas (ed.), *Growing Greener Cities in Latin America and the Caribbean: An FAO Report on Urban and Peri-urban Agriculture in the Region*, pp. 10–19. Rome: FAO.

Goode, Erica. 2015. 'Cuba's Environmental Concerns Grow with Prospect of U.S. Presence'. *New York Times*, 1 July. www.nytimes.com/2015/07/02/science/earth/ cubas-environmental-concerns-grow-with-prospect-of-us-presence.html (accessed 24 October 2020).

Gore, Tim. 2020. 'Confronting Carbon Inequality: Putting Climate Justice at the Heart of the COVID-19 Recovery'. https://oxfamilibrary.openrepository.com/ bitstream/handle/10546/ 621052/mb-confronting-carbon-inequality-210920-en. pdf (accessed 9 October 2020).

Graham, Helen. 2002. *The Spanish Republic at War 1936–1939*. Cambridge: Cambridge University Press.

Graham, Robert, ed. 2005. *Anarchism: A Documentary History of Libertarian Ideas. Volume 1: From Anarchy to Anarchism (300 CE to 1939)*. Montréal: Black Rose Books.

Gray, Paul C., ed. 2018. *From the Streets to the State: Changing the World by Taking Power*. Albany: State University of New York Press.

Gregorich, Edward G., and Darwin W. Anderson. 1985. 'Effects of Cultivation and Erosion on Soils of Four Toposequences in the Canadian Prairies'. *Geoderma* 36 (3–4): 343–54.

Grennfelt, Peringe, Anna Engleryd, Martin Forsius, Øystein Hov, Henning Rodhe and Ellis Cowling. 2020. 'Acid Rain and Air Pollution: 50 Years of Progress in Environmental Science and Policy'. *Ambio* 49: 849–64.

Griffiths, Patrick, Daniel Müller, Tobias Kuemmerle and Patrick Hostert. 2013. 'Agricultural Land Change in the Carpathian Ecoregion after the Breakdown of Socialism and Expansion of the European Union'. *Environmental Research Letters* 8: 045024.

Griffiths, Patrick, Tobias Kuemmerle, Matthias Baumann, Volker C. Radeloff, Ioan V. Abrudan, Juraj Lieskovsky, Catalina Munteanu, Katarzyna Ostapowicz and Patrick Hostert. 2014. 'Forest Disturbances, Forest Recovery, and Changes in Forest Types across the Carpathian Ecoregion from 1985 to 2010 Based on Landsat Image Composites'. *Remote Sensing of Environment* 151: 72–88.

Guérin, Daniel. 1970. *Anarchism: From Theory to Practice*. New York: Monthly Review.

Gupta, Ramesh C., and Daya R. Varma. 2020. 'Methyl Isocyanate: The Bhopal Gas', in Ramesh C. Gupta (ed.), *Handbook of Toxicology of Chemical Warfare Agents* (3rd edition), pp. 389–402. Cambridge: Academic Press.

Hartman, Chester, and Gregory Squires. 2006. *There Is No Such Thing as a Natural Disaster: Race, Class, and Hurricane Katrina*. New York: Routledge.

Harvey, David. 1996. *Justice, Nature and the Geography of Difference*. Oxford: Blackwell.

Harvey, Fiona. 2020. 'Half UK's True Carbon Footprint Created Abroad, Research Finds'. *The Guardian*, 16 April, www.theguardian.com/environment/2020/apr/16/britain-climate-efforts-undermined-failure-imports-carbon, accessed 21 October 2020.

Heller, Henry. 2011. *The Birth of Capitalism: A 21st Century Perspective*. London: Pluto Press.

Herring, Stephanie C., Nikolaos Christidis, Andrew Hoell, Martin P. Hoerling and Peter A. Stott, 2020. 'Explaining Extreme Events of 2018 from a Climate Perspective'. *Bulletin of the American Meteorological Society* 101: S1–S140.

Hill, Malcolm R. 1997. *Environment and Technology in the Former USSR: The Case of Acid Rain and Power Generation*. Cheltenham: Edward Elgar.

Hongtao Liu, Youmin Xi, Ju'e Guo and Xia Li. 2010. 'Energy Embodied in the International Trade of China: An Energy Input–Output Analysis'. *Energy Policy* 38 (8): 3957–64.

Hornborg, Alf, Gustav Cederlöf and Andreas Roos. 2019. 'Has Cuba Exposed the Myth of "Free" Solar Power? Energy, Space, and Justice'. *Environment and Planning E: Nature and Space* 2 (4): 989–1008.

Horne, Gerald. 1986. *Black and Red: W.E.B Du Bois and the Afro-American Response to the Cold War, 1944–1963*. Albany: SUNY Press.

Horowitz, Howard. 2012. 'Aral Sea Analogs in the American West', in Michael R. Edelstein, Astrid Cerny and Abror Gadaev (eds), *Disaster by Design: The Aral Sea and Its Lessons for Sustainability*, pp. 89–104. Bingley: Emerald Group.

Houghton, Richard A. 2016. 'Deforestation', in John F. Shroder and Ramesh Sivanpillai (eds), *Biological and Environmental Hazards, Risks, and Disasters*, pp. 313–15. New York: Academic Press.

Huan, Qingzhi. 2016. 'Socialist Eco-civilization and Social-Ecological Transformation'. *Capitalism Nature Socialism* 27 (2): 51–66.

Huang, Phillip C.C. 2012. 'Profit-Making State Firms and China's Development Experience: "State Capitalism" or "Socialist Market Economy"?' *Modern China* 38 (6): 591–629.

Huang, Rui, Manfred Lenzen and Arunima Malik. 2019. 'CO_2 Emissions Embodied in China's Export'. *The Journal of International Trade & Economic Development* 28 (8): 919–34.

Huberman, Leo, and Paul M. Sweezy. 1968. *Introduction to Socialism*. New York: Monthly Review.

Huffman, E., R.G. Eilers, G. Padbury, G. Wall and K.B. MacDonald. 2000. 'Canadian Agri-environmental Indicators Related to Land Quality: Integrating Census and Biophysical Data to Estimate Soil Cover, Wind Erosion and Soil Salinity'. *Agriculture, Ecosystems & Environment* 81 (2): 113–23.

Hurley, Andrew. 1988. 'The Social Biases of Environmental Change in Gary, Indiana, 1945–1980'. *Environmental Review* 12 (4): 1–19.

References

Iakovleva, Emiliia V., Daniel D. Guerra and Andrey Y. Shklyarskiy. 2020. 'Alternative Measures to Reduce Carbon Dioxide Emissions in the Republic of Cuba'. *Journal of Ecological Engineering* 21 (4): 55–60.

Igreja, Victor. 2010. 'Frelimo's Political Ruling through Violence and Memory in Postcolonial Mozambique'. *Journal of Southern African Studies* 36 (4): 781–99.

INIFAT (Instituto de Investigaciones Fundamentales en Agricultura Tropical). 2010. *Manual Técnico para Organopónicos, Huertos Intensivos y Organoponía Semiprotegida* (7th edition). Havana: INIFAT.

Ipingbemi, Olusiyi. 2009. 'Socio-economic Implications and Environmental Effects of Oil Spillage in Some Communities in the Niger Delta'. *Journal of Integrative Environmental Sciences* 6 (1): 7–23.

Iwegbue, Chukwujindu M.A., E.S. Williams and N.O. Isirimah. 2009. 'Study of Heavy Metal Distribution in Soils Impacted with Crude Oil in Southern Nigeria'. *Soil & Sediment Contamination* 18: 136–43.

Izmerov, Nikolaj Fedotovic. 1973. *Control of Air Pollution in the USSR*. Geneva: World Health Organization.

Jancar, Barbara. 1987. *Environmental Management in the Soviet Union and Yugoslavia: Structure and Regulation in Federal Communist Systems*. Durham, NC: Duke University Press.

Jing, Jun. 2010. 'Environmental Protests in Rural China', in Elizabeth J. Perry and Mark Selden (eds), *Chinese Society: Change, Conflict and Resistance* (3rd edition), pp. 197–214. London: Routledge.

Johnson, Sapna, Ramakant Satu, Nimisha Jadon and Clara Duca. 2009. *Contamination of Soil and Water inside and outside the Union Carbide India Limited, Bhopal*. New Delhi: Centre for Science and Environment.

Josephson, Paul, Nicolai Dronin, Aleh Cherp, Ruben Mnatsakanian, Dmitry Efremenko and Vladislav Larin. 2013. *An Environmental History of Russia*. Cambridge: Cambridge University Press.

Kachur, Anatoly N., Valentina S. Arzhanova, Pavel V. Yelpatyevsky, Margrit C. von Braun and Ian H. von Lindern. 2003. 'Environmental Conditions in the Rudnaya River Watershed: A Compilation of Soviet and Post-Soviet Era Sampling around a Lead Smelter in the Russian Far East'. *Science of The Total Environment* 303 (1–2): 171–85.

Kalu, Victoria, and Ngozi Stewart. 2007. 'Nigeria's Niger Delta Crises and Resolution of Oil and Gas Related Disputes: Need for a Paradigm Shift'. *Journal of Energy & Natural Resources Law* 25 (3): 244–67.

Kan, Haidong. 2009. 'Environment and Health in China: Challenges and Opportunities'. *Environmental Health Perspectives* 117 (12): A530–A531.

Kanianska, Radoslava, Miriam Kizeková, Jozef Nováček and Martin Zeman. 2014. 'Land-Use and Land-Cover Changes in Rural Areas during Different Political Systems: A Case Study of Slovakia from 1782 to 2006'. *Land Use Policy* 36: 554–66.

Karimov, Bakhtiyor. 2011. 'An Overview on Desert Aquaculture in Central Asia (Aral Sea Drainage Basin)', in Valerio Crespi and Alessandro Lovatelli (eds),

Aquaculture in Desert and Arid Lands: Development Constraints and Opportunities. FAO Technical Workshop, 6–9 July 2010, Hermosillo, Mexico. *FAO Fisheries and Aquaculture Proceedings* No. 20, pp. 61–84. Rome: FAO.

Kariyeva, Jahan, and Willem J.D. van Leeuwen. 2012. 'Phenological Dynamics of Irrigated and Natural Drylands in Central Asia before and after the USSR Collapse'. *Agriculture, Ecosystems & Environment* 162: 77–89.

Kasar, Sharayu, Suchismita Mishra, Yasutaka Omori, Sarata Kumar Sahoo, Norbert Kavasi, Hideki Arae, Atsuyuki Sorimachi and Tatsuo Aono. 2020. 'Sorption and Desorption Studies of Cs and Sr in Contaminated Soil Samples around Fukushima Daiichi Nuclear Power Plant'. *Journal of Soils Sediments* 20: 392–403.

Kelley, Donald R. 1976. 'Environmental Policy-Making in the USSR: The Role of Industrial and Environmental Interest Groups'. *Soviet Studies* 28 (4): 570–89.

Kelley, Donald R., Kennether R. Stunkel and Richard R. Wescott. 1976. *The Economic Superpowers and the Environment: The United States, the Soviet Union, and Japan*. San Francisco: W.H. Freeman.

Kinkade, Kat, with the Twin Oaks Community. 2011. 'Labour Credit: Twin Oaks Community', in Anitra Nelson and Frans Timmerman (eds), *Life without Money: Building Fair and Sustainable Economies*, pp. 173–91. London: Pluto Press.

Komarov, Boris. 1980 [1978]. *The Destruction of Nature in the Soviet Union*. London: M.E. Sharpe.

Koont, S. 2009. 'The Urban Agriculture of Havana'. *Monthly Review* 60 (8): 44–63.

Koont, S. 2011. *Sustainable Urban Agriculture in Cuba*. Gainesville: University Press of Florida.

Korovin, N.V. 1994. 'Development of Fuel Cells and Electrolyzers in the Former U.S.S.R'. *International Journal of Hydrogen Energy* 19 (9): 771–6.

Kovel, Joel. 2007. *The Enemy of Nature: The End of Capitalism or the End of the World?* London: Zed Books.

Kovel, Joel. 2014. 'Ecosocialism as a Human Phenomenon'. *Capitalism Nature Socialism* 25 (1): 10–23.

Kowalewski, Michal, Guillermo Serrano, Karl Flessa and Glenn Goodfriend. 2000. 'Dead Delta's Former Productivity: Two Trillion Shells at the Mouth of the Colorado River'. *Geology* 28: 1059–62.

Kozlov, Mikhail V., and Valery Barcan. 2000. 'Environmental Contamination in the Central Part of the Kola Peninsula: History, Documentation, and Perception'. *AMBIO: A Journal of the Human Environment* 29 (8): 512–17.

Krader, Lawrence. 1974. *The Ethnological Notebooks of Karl Marx*. Assen: Van Gorcum.

Kraemer, Roland, Alexander V. Prishchepov, Daniel Müller, Tobias Kuemmerle, Volker C. Radeloff, Andrey Dara, Alexey Terekhov and Manfred Frühauf. 2015. 'Long-Term Agricultural Land-Cover Change and Potential for Cropland Expansion in the Former Virgin Lands Area of Kazakhstan'. *Environmental Research Letters* 10: 054012. https://iopscience.iop.org/article/10.1088/1748-9326/10/5/054012/pdf.

References

Krausmann, Fridolin, Birgit Gaugl, James West and Heinz Schandl. 2016. 'The Metabolic Transition of a Planned Economy: Material Flows in the USSR and the Russian Federation 1900 to 2010'. *Ecological Economics* 124: 76–85.

Krüger, Olaf, Roman Marks and Hartmut Graßl. 2004. 'Influence of Pollution on Cloud Reflectance'. *Journal of Geophysical Research* 109 (D24). https://doi.org/10.1029/2004JD004625.

Kuemmerle, Tobias, Daniel Müller, Patrick Griffiths and Marioara Rusu. 2009. 'Land Use Change in Southern Romania after the Collapse of Socialism'. *Regional Environmental Change* 9: 1. https://doi.org/10.1007/s10113-008-0050-z.

Kuznetsova, V.I. 2012. 'Changes in the Physical Status of the Typical and Leached Chernozems of Kursk Oblast within 40 Years'. *Eurasian Soil Science* 46 (4): 393–400.

Laity, Julie. 2008. *Deserts and Desert Environments*. Chichester: Wiley.

Lane, David. 2014. *The Capitalist Transformation of State Socialism: The Making and Breaking of State Socialist Society, and What Followed*. London: Routledge.

Leitgeb, Friedrich, Reinaldo Funes Monzote, Susanne Kummer and Christian R. Vogl. 2011. 'Contribution of Farmers' Experiments and Innovations to Cuba's Agricultural Innovation System'. *Renewable Agriculture and Food Systems* 26 (4): 354–67.

Leitgeb, Friedrich, Sarah Schneider and Christian R. Vogl. 2016. 'Increasing Food Sovereignty with Urban Agriculture in Cuba'. *Agriculture and Human Values* 33: 415–26.

Lerman, Zvi, and Astghik Mirzakhanian. 2001. *Private Agriculture in Armenia*. Lanham: Lexington Books.

Levins, Richard. 1990. 'The Struggle for Ecological Agriculture in Cuba'. *Capitalism Nature Socialism* 1 (5): 121–41.

Levins, Richard. 2004. 'Cuba's Biological Weapons'. *Capitalism Nature Socialism* 15 (2): 31–3.

Levins, Richard. 2005. 'How Cuba Is Going Ecological'. *Capitalism Nature Socialism* 16 (3): 7–25.

Levins, Richard, and Richard Lewontin. 1985. *The Dialectical Biologist*. Boston: Harvard University Press.

Li, Hong, Zhang Pei Dong, He Chunyu and Wang Gang. 2007. 'Evaluating the Effects of Embodied Energy in International Trade on Ecological Footprint in China'. *Ecological Economics* 62 (1): 136–48.

Li, Minqi. 2009. *The Rise of China and the Demise of the Capitalist World Economy*. New York: Monthly Review Press.

Li, Minqi. 2016. *China and the 21st Century Crisis*. London: Pluto Press.

Li, Minqi. 2017. 'Political Economy of China from Socialism to Capitalism, and to Eco-Socialism?' in Robert Westra (ed.), *The Political Economy of Emerging Markets Varieties of BRICS in the Age of Global Crises and Austerity*, pp. 79–100. London: Routledge.

Li, Tiankui, Yi Liu, Sijie Lin, Yangze Liu and Yunfeng Xie. 2019. 'Soil Pollution Management in China: A Brief Introduction'. *Sustainability* 11: 556.

Lieskovský, Juraj, Róbert Kanka, Peter Bezák, Dagmar Štefunková, František Petrovič and Marta Dobrovodská. 2013. 'Driving Forces behind Vineyard Abandonment in Slovakia Following the Move to a Market-Oriented Economy'. *Land Use Policy* 32: 356–65.

Lin, Boqiang, and Chuanwang Sun. 2010. 'Evaluating Carbon Dioxide Emissions in International Trade of China'. *Energy Policy* 38 (1): 613–21.

Lin, Jintai, Da Pan, Steven J. Davis, Qiang Zhang, Kebin He, Can Wang, David G. Streets, Donald J. Wuebbles and Dabo Guan. 2014. 'Trade and Transport of Air Pollution'. *Proceedings of the National Academy of Sciences* 111 (5): 1736–41.

Lin, Jintai, Mingxi Du, Lulu Chen, Kuishuang Feng, Yu Liu, Randall V. Martin, Jingxu Wang, Ruijing Ni, Yu Zhao, Hao Kong, Hongjian Weng, Mengyao Liu, Aaron van Donkelaar, Qiuyu Liu and Klaus Hubacek. 2019. 'Carbon and Health Implications of Trade Restrictions'. *Nature Communications* 10: 4947.

Lindert, Peter H. 2000. *Shifting Ground: The Changing Agricultural Soils of China and Indonesia*. Cambridge, MA: MIT Press.

Lioubimtseva, Elena. 2014. 'Impact of Climate Change on the Aral Sea and Its Basin', in Philip Macklin (ed.), *The Aral Sea*, pp. 405–27. Berlin: Springer.

Litvinov, N. 1962. 'Water Pollution in the USSR and Other Eastern European Countries'. *Bulletin of the World Health Organization* 26 (4): 439–63.

Liu, Mingliang, and Hanqin Tian. 2010. 'China's Land Cover and Land Use Change from 1700 to 2005: Estimations from High-Resolution Satellite Data and Historical Archives'. *Global Biogeochemical Cycles* 24: GB3003.

Liu, Jianguo, and Jared Diamond. 2005. 'China's Environment in a Globalizing World'. *Nature* 435: 1179–86.

Long, Zhiming, Rémy Herrera and Tony Andréani. 2018. 'On the Nature of the Chinese Economic System'. *Monthly Review* 70: 32–43.

Losurdo, Domenico. 2005. *Controstoria del Liberalismo* [Liberalism: A Counter-History.] Rome: Editori Laterza.

Lowell, Jonathan T., and Sara Law. 2017. 'Sustainability's Incomplete Circles: Towards a Just Food Politics in Austin, Texas and Havana, Cuba', in Antoinette M.G.A. WinklerPrins (ed.), *Global Urban Agriculture: Convergence of Theory and Practice between North and South*, pp. 106–17. Boston: CABI.

Löwy, Michael. 2011. *Écosocialisme: L'Alternative Radicale à la Catastrophe Écologique Capitaliste*. [Ecoscialism: The Radical Alternative to Capitalist Ecological Catastrophe.] Paris: Fayard.

Löwy, Michael. 2017. 'Marx, Engels, and Ecology'. *Capitalism Nature Socialism* 28 (2): 10–21.

Lucantoni, Dario. 2020. 'Transition to Agroecology for Improved Food Security and Better Living Conditions: Case Study from a Family Farm in Pinar del Río, Cuba'. *Agroecology and Sustainable Food Systems* 44 (9): 1124–61.

References

Maas, Rob, Peringe Grennfelt, Markus Amann, Bill Harnett, Jennifer Kerr, Eva Berton, Dominique Pritula, Ilze Reiss, Paul Almodovar, Marie-Eve Héroux, David Fowler, Dick Wright, Heleen de Wit, Kjetil Tørseth, Katja Mareckova, Anne-Christine LeGall, Isaura Rabago, Jean-Paul Hettelingh, Richard Haeuber and Stefan Reis. 2016. *Towards Cleaner Air: Scientific Assessment Report 2016 – EMEP Steering Body and Working Group on Effects of the Convention on Long-Range Transboundary Air Pollution*. Oslo: UNECE.

Machado, Mario Reynaldo. 2017. 'Alternative to What? Agroecology, Food Sovereignty, and Cuba's Agricultural Revolution'. *Human Geography* 10 (3): 7–21.

Machín Sosa, Braulio, Adilén María Roque Jaime, Dana Rocío Ávila Lozano and Peter Michael Rosset (2013). *Agroecological Revolution: The Farmer-to-Farmer Movement of the ANAP in Cuba*. Havana: ANAP and Via Campesina.

Madrazo, Jessie, Alain Clappier, Luis Belalcazar, Osvaldo Cuesta, Heydi Contreras and François Golay. 2018. 'Screening Differences between a Local Inventory and the Emissions Database for Global Atmospheric Research (EDGAR)'. *Science of the Total Environment* 631–2: 934–41.

Malm, Andreas. 2012. 'China as Chimney of the World: The Fossil Capital Hypothesis'. *Organization & Environment* 25 (2): 146–77.

Manji, Firoze, and Bill Fletcher, eds. 2013. *Claim No Easy Victories: The Legacy of Amílcar Cabral*. Dakar: CODESRIA and Daraja Press.

Marks, Robert B. 1998. *Tigers, Rice, Silk, and Silt: Environment and Economy in Late Imperial South China*. Cambridge: Cambridge University Press.

Marks, Robert B. 2017. *China: An Environmental History*. Lanham: Rowman & Littlefield.

Marshalek, Frank. 2017. 'Cuban and Danish Agriculture, the Rochdale Principles, and the Renovation of Socialism'. *Human Geography* 10 (3): 22–40.

Martz, L.W., and E. de Jong. 1987. 'Using Cesium-137 to Assess the Variability of Net Soil Erosion and Its Association with Topography in a Canadian Prairie Landscape'. *CATENA* 14 (5): 439–51.

Marx, Karl. 1992 [1867]. *Capital: A Critical Analysis of Capitalist Production, Volume 1*. Translated by S. Moore and E. Aveling. New York: International Publishers.

McCauley, Martin. 1976. *Khrushchev and the Development of Soviet Agriculture: The Virgin Land Programme 1953–1964*. New York: Holmes & Meier.

McIntyre, Robert J., and James R. Thornton. 1978. 'On the Environmental Efficiency of Economic Systems'. *Soviet Studies* 30 (2): 173–92.

Meisenhelder, Thomas. 1995. 'Marx, Engels, and Africa'. *Science & Society* 59: 197–205.

Mejia, Steven A. 2020. 'Foreign Direct Investment Dependence and the Neglected Greenhouse Gas: A Cross-national Analysis of Nitrous Oxide Emissions in Developing Countries, 1990–2014'. *Sociological Perspectives*. https://doi.org/10.1177/0731121420937738.

Meneses, Maria P. 2016. 'Hidden Processes of Reconciliation in Mozambique: The Entangled Histories of Truth-Seeking Commissions Held between 1975 and 1982'. *Africa Development/Afrique Et Développement* 41 (4): 153–80.

Merchant, Carolyn. 1980. *The Death of Nature: Women, Ecology, and the Scientific Revolution*. San Francisco: Harper.

Micklin, Philip. 2007. 'The Aral Sea Disaster'. *Annual Review of Earth and Planetary Sciences* 35: 47–72.

Mies, Maria. 1986. *Patriarchy and Accumulation on a World Scale*. London: Zed Books.

Mignon Kirchhof, Astrid, and John R. McNeill, eds. 2019. 'Introduction', in Astrid Mignon Kirchhof and John R. McNeill (eds), *Nature and the Iron Curtain: Environmental Policy and Social Movements in Communist and Capitalist Countries, 1945–1990*, pp. 3–14. Pittsburgh: University of Pittsburgh Press.

Miliband, Ralph. 1994. *Socialism for a Sceptical Age*. Cambridge: Polity Press.

Milner-Gulland, E.J., O.M. Bukreevea, T. Coulson, A.A. Lushchekina, M.V. Kholodova, A.B. Bekenov and I.A. Grachev. 2003. 'Conservation: Reproductive Collapse in Saiga Antelope Harems'. *Nature* 422: 135–5.

Mishra, Pradyumna K., Ravindram M. Samarth, Neelam Pathak, Subodh K. Jain, Smita Banerjee and Kewal K. Maudar. 2009. 'Bhopal Gas Tragedy: Review of Clinical and Experimental Findings after 25 Years'. *International Journal of Occupational Medicine and Environmental Health* 22 (3): 193–202.

Mitman, Gregg, Michelle Murphy and Christopher Sellers. 2004. 'Introduction: A Cloud over History'. *Osiris* 19: 1–17.

Mittal, Alok. 2016. 'Retrospection of Bhopal Gas Tragedy'. *Toxicological & Environmental Chemistry* 98 (9): 1079–83.

Montgomery, David R. 2007. *Dirt: The Erosion of Civilizations*. Berkeley: University of California Press.

Monzote, Reinaldo Funes. 2008. *From Rainforest to Cane Field in Cuba: An Environmental History since 1492*. Translated by A. Martin. Raleigh: The University of North Carolina Press.

Moran, Daniel D., Mathis Wackernagel, Justin A. Kitzes, Steven H. Goldfinger and Aurélien Boutaud. 2008. 'Measuring Sustainable Development: Nation by Nation'. *Ecological Economics* 64 (3): 470–4.

Mueller, Gustav E. 1958. 'The Hegel Legend of "Thesis-Antithesis-Synthesis"'. *Journal of the History of Ideas* 19 (3): 411–14.

Muldavin Joshua. 2000. 'The Paradoxes of Environmental Policy and Resource Management in Reform-Era China'. *Economic Geography* 76 (3): 244–71.

Munteanu, Catalina, Tobias Kuemmerle, Martin Boltiziar, Van Butsic, Urs Gimmi, Lúboš Halada, Dominik Kaim, Géza Király, Éva Konkoly-Gyuró, Jacek Kozak, Juraj Lieskovský, Matej Mojses, Daniel Müller, Krzystof Ostafin, Katarzyna Ostapowicz, Oleksandra Shandra, Přemysl Štych, Sarah Walker and Volker C. Radeloff. 2014. 'Forest and Agricultural Land Change in the Carpathian Region: A Meta-analysis of Long-Term Patterns and Drivers of Change'. *Land Use Policy* 38: 685–97.

NASA. 2012. 'Landsat Top Ten: A Shrinking Sea, Aral Sea'. www.nasa.gov/mission_pages/landsat/news/40th-top10-aralsea.html (accessed 15 November 2020).

References

Navarro, Vicente. 1992. 'Has Socialism Failed? An Analysis of Health Indicators under Socialism'. *International Journal of Health Services* 22 (4): 583–601.

Nelson, Anitra. 2018. *Small Is Necessary: Shared Living on a Shared Planet*. London: Pluto Press.

Newell, Joshua P., and Laura A. Henry. 2017. 'The State of Environmental Protection in the Russian Federation: A Review of the Post-Soviet Era'. *Eurasian Geography and Economics* 57 (6): 779–801.

Nijman, Vincent. 2010. 'An Overview of International Wildlife Trade from Southeast Asia'. *Biodiversity and Conservation* 19: 1101–14.

O'Connor, James. 1970. *The Origins of Socialism in Cuba*. Ithaca, NY: Cornell University Press.

O'Connor, James. 1988. 'Prospectus. Capitalism Nature Socialism: A Journal of Socialist Ecology'. *Capitalism Nature Socialism* 1 (1): 1–6.

O'Connor, James. 1991. 'Socialism and Ecology'. *Capitalism Nature Socialism* 2 (3): 1–12.

O'Connor, James. 1998. *Natural Causes: Essays in Ecological Marxism*. New York: Guilford.

Obertreis, Julia. 2018. 'Soviet Irrigation Policies under Fire: Ecological Critique in Central Asia, 1970s–1991'. In Nicholas Breyfogle (ed.), *Eurasian Environments: Nature and Ecology in Eurasian History*, pp. 113–29. Pittsburgh: University of Pittsburgh Press.

Oldfield, Jonathan. 2005. *Russian Nature: Exploring the Environmental Consequences of Societal Change*. New York: Routledge.

Oldfield, Jonathan. 2018. 'Imagining Climates Past, Present and Future: Soviet Contributions to the Science of Anthropogenic Climate Change, 1953–1991'. *Journal of Historical Geography* 60: 41–51.

Olivares-Rieumont, Susana, Daniel de la Rosa, Lazaro Lima, David W. Graham, Katia D'Alessandro, Jorge Borroto, Francisco Martínez and J. Sánchez. 2005. 'Assessment of Heavy Metal Levels in Almendares River Sediments: Havana City, Cuba'. *Water Research* 39 (16): 3945–53.

Ostergren, David, and Evgeny Shvarts. 2000. 'Russian Zapovedniki in 1998: Recent Progress and New Challenges for Russia's Strict Nature Preserves'. *USDA Forest Service Proceedings* RMRS-P-14: 209–13.

Ostergren, David, and Peter Jacques. 2002. 'A Political Economy of Russian Nature Conservation Policy: Why Scientists Have Taken a Back Seat'. *Global Environmental Politics* 2 (4): 102–24.

Pavlínek, Petr, and John Pickles. 2000. *Environmental Transitions: Transformation and Ecological Defence in Central and Eastern Europe*. New York: Routledge.

Permitin, Viktor E., and Vladimir S. Tikunov. 1992. 'Environmental Monitoring in the USSR: Present State and New Tasks'. *International Journal of Environmental Studies* 40 (1): 67–77.

Perrault, Gilles, ed. 1998. *Le Livre Noir du Capitalisme* [The Black Book of Capitalism.] Paris: Le Temps des Cerises.

Peters, Glenn, Greg Marland, Corinne Le Quéré, Thomas Boden, Josep G. Canadell and Michael R. Raupach. 2012. 'Rapid Growth in CO_2 Emissions after the 2008–2009 Global Financial Crisis'. *Nature Climate Change* 2: 2–4.

Peterson, D.J. 1993. *Troubled Lands: The Legacy of Soviet Environmental Destruction.* A RAND Research Study. Boulder: Westview Press.

Petr, K., K. Ismukhanov, B. Kamilov, D. Pulatkhon and P.D. Umarov. 2004. 'Irrigation Systems and Their Fisheries in the Aral Sea Basin, Central Asia', in R. Welcomme and T. Petr (eds), *Proceedings of the Second International Symposium on the Management of Large Rivers for Fisheries Volume II*, FAO Regional Office for Asia and the Pacific, Bangkok, Thailand. RAP Publication 2004/17. www.fao.org/3/ad526eoi.htm (accessed 15 November 2020).

Placeres, Manuel R., Maricel G. Melián and Mireya A. Toste. 2011. 'Principales Características de la Salud Ambiental de la Provincia La Habana'. [Main Characteristics of Environmental Health in Habana Province.] *Revista Cubana de Higiene y Epidemiología* 49 (3): 384–98.

Plumwood, Val. 1993. *Feminism and the Mastery of Nature*. New York: Routledge.

Pomeranz, Kenneth. 2009a. 'Introduction: World History and Environmental History', in Edmund Burke III and Kenneth Pomeranz (eds), *The Environment and World History*, pp. 3–32. Berkeley: University of California Press.

Pomeranz, Kenneth. 2009b. 'The Great Himalayan Watershed: Agrarian Crisis, Mega-dams and the Environment'. *New Left Review* 58: 5–39.

Potapov, P.V., S.A. Turubanova, A. Tyukavina, A.M. Krylov, J.L. McCarty, V.C. Radeloff and M.C. Hansen. 2015. 'Eastern Europe's Forest Cover Dynamics from 1985 to 2012 Quantified from the Full Landsat Archive'. *Remote Sensing of Environment* 159: 28–43.

Prashad, Vijay. 2018. *Red Star over the Third World*. New Delhi: LeftWord.

Premat, Adriana. 2012. *Sowing Change: The Making of Urban Agriculture in Havana.* Nashville: Vanderbilt University Press.

Pryde, Philip R. 1972. *Conservation in the Soviet Union*. Cambridge: Cambridge University Press.

Pryde, Philip R. 1986. 'Strategies and Problems of Wildlife Preservation in the USSR'. *Biological Conservation* 36: 351–74.

Pryde, Philip R. 1991. *Environmental Management in the Soviet Union*. Cambridge: Cambridge University Press.

Ramanathan, V., C. Chung, D. Kim, T. Bettge, L. Buja, J. T. Kiehl, W. M. Washington, Q. Fu, D. R. Sikka and M. Wild. 2005. 'Atmospheric Brown Clouds: Impacts on South Asian Climate and Hydrological Cycle'. *Proceedings of the National Academy of Sciences* 102 (15): 5326–33.

Ravishankara, A.R., John S. Daniel and Robert W. Portmann. 2009. 'Nitrous Oxide (N_2O): The Dominant Ozone-Depleting Substance Emitted in the 21st Century'. *Science* 326 (5949): 123–5.

References

Reclus, Elisée. 2013. *Anarchy, Geography, Modernity: Selected Writings of Elisée Reclus*. Introduced and edited by Camille Martin and John Clark. Oakland: PM Press.

Ren, Shenggang, Baolong Yuan, Xie Ma and Xiaohong Chen. 2014. 'The Impact of International Trade on China's Industrial Carbon Emissions since Its Entry into WTO'. *Energy Policy* 69: 624–34.

Resnick, Stephen A., and Richard D. Wolff. 2002. *Class Theory and History: Capitalism and Communism in the USSR*. New York: Routledge.

Revich, Boris A. 1995. 'Public Health and Ambient Air Pollution in Arctic and Subarctic Cities of Russia'. *The Science of the Total Environment* 160–1: 585–92.

Rim-Rukeh, A., G. Ikhifa and P. Okokoyo. 2007. 'Physico-chemical Characteristics of Some Waters Used for Drinking and Domestic Purposes in the Niger Delta, Nigeria'. *Environmental Monitoring & Assessment* 128 (1–3): 475–82.

Ripple, William J., Christopher Wolf, Thomas M. Newsome, Phoebe Barnard and William R. Moomaw. 2020. 'World Scientists' Warning of a Climate Emergency'. *BioScience* 70 (1): 8–12.

Ritchie, Hannah, and Max Roser. 2017. 'CO_2 and Greenhouse Gas Emissions'. https://ourworldindata.org/co2-and-other-greenhouse-gas-emissions (accessed 21 October 2020).

Rizo, Oscar Díaz, M. Hernández Merlo, Frank Echeverría Castillo and J. Arado López. 2012. 'Assessment of Metal Pollution in Soils from a Former Havana (Cuba) Solid Waste Open Dump'. *Bulletin of Environmental Contamination & Toxicology* 88 (2): 182–6.

Rizo, Oscar Díaz, Frank Echeverría Castillo, J. Arado López, J. and M. Hernández Merlo. 2011. 'Assessment of Heavy Metal Pollution in Urban Soils of Havana City, Cuba'. *Bulletin of Environmental Contamination & Toxicology* 87 (4): 414–19.

Roberts, J. Timmons, Peter E. Grimes and Jodie L. Manale. 2003. 'Social Roots of Global Environmental Change: A World-Systems Analysis of Carbon Dioxide Emissions'. *Journal of World-Systems Research* 9 (2): 277–315.

Robinson, Michael C. 1989. *The Mississippi River Commission: An American Epic*. Vicksburg: Mississippi River Commission, U.S. Army Corps of Engineers.

Robinson, Michael E. 2007. *Korea's Twentieth-Century Odyssey: A Short History*. Honolulu: University of Hawai'i Press.

Robinson Sarah. 2016. 'Land Degradation in Central Asia: Evidence, Perception and Policy', in Roy Behnke and Michael Mortimore (eds), *The End of Desertification?*, pp. 451–90. Berlin: Springer.

Rockhill, Gabriel. 2017. *Counter-History of the Present: Untimely Interrogations into Globalization, Technology, Democracy*. Durham, NC: Duke University Press.

Rodney, Walter. 2011 [1972]. *How Europe Underdeveloped Africa*. Baltimore: Black Classic Press.

Roman, Joe. 2018. 'The Ecology and Conservation of Cuba's Coastal and Marine Ecosystems'. *Bulletin of Marine Science* 94 (2): 149–69.

Rosset, Peter, and Medea Benjamin. 1994. 'Cuba's Nationwide Conversion to Organic Agriculture'. *Capitalism Nature Socialism* 5 (3): 79–97.

Rosset, Peter Michael, Braulio Machín Sosa, Adilén María Roque Jaime and Dana Rocío Ávila Lozano. 2011. 'The Campesino-to-Campesino Agroecology Movement of ANAP in Cuba: Social Process Methodology in the Construction of Sustainable Peasant Agriculture and Food Sovereignty'. *The Journal of Peasant Studies* 38 (1): 161–91.

Royce, Frederick. 2018. 'Agricultural Production Co-operatives in Cuba: Toward Sustainability', in Sonja Novković and Henry Veltmeyer (eds), *Co-operativism and Local Development in Cuba: An Agenda for Democratic Social Change*, pp. 128–59. Leiden: Brill.

San Sebastián, Miguel, Benedict Armstrong, Juan A. Córdoba and Carolyn Stephens. 2001. 'Exposures and Cancer Incidence near Oil Fields in the Amazon Basin of Ecuador'. *Occupational and Environmental Medicine* 58: 517–22.

Sánchez-Bayo, Francisco, and Kris A.G. Wyckhuys. 2019. 'Worldwide Decline of the Entomofauna: A Review of Its Drivers'. *Biological Conservation* 232: 8–27.

Sanchez-Sibony, Oscar. 2014. *Red Globalization: The Political Economy of the Soviet Cold War from Stalin to Khrushchev.* Cambridge: Cambridge University Press.

Sankara, Thomas. 2007 [1986]. *Thomas Sankara Speaks: The Burkina Faso Revolution 1983–87* (2nd edition). Edited by Michael Prairie. New York: Pathfinder Press.

Sarkodie, Samuel Asumadu, Samuel Adams and Thomas Leirvik. 2020. 'Foreign Direct Investment and Renewable Energy in Climate Change Mitigation: Does Governance Matter?' *Journal of Cleaner Production* 263: 121262.

Scarpaci, Joseph L., and Armando H. Portela. 2005. 'The Historical Geography of Cuba's Sugar Landscape', in Jorge F. Pérez-López and José Álvarez (eds), *Reinventing the Cuban Sugar Agroindustry*, pp. 11–25. Lanham: Lexington Books.

Schecter, Arnold, Le C. Dai, Le Thi B. Thuy, Hoan T. Quynh, Ding Q. Minh, Hoang D. Cau, Pham H. Phiet, N.T. Nguyen, John D. Constable and Robert Baughman. 1995. 'Agent Orange and the Vietnamese: The Persistence of Elevated Dioxin Levels in Human Tissues'. *American Journal of Public Health* 85: 516–22.

Schneiders, Berts. 1996. 'The Myth of Environmental Management: The Corps, the Missouri River, and the Channelization Project'. *Agricultural History* 70 (2): 337–50.

Schwartzman, David. 1996. 'Solar Communism'. *Science & Society* 60 (3): 307–31.

Schwartzman, David, and Quincy Saul. 2015. 'An Ecosocialist Horizon for Venezuela: A Solar Communist Horizon for the World'. *Capitalism Nature Socialism* 26: 14–30.

Schwartzman, Peter, and David Schwartzman. 2019. *The Earth Is Not for Sale: A Path Out of Fossil Capitalism to the Other World that Is Still Possible.* Singapore: World Scientific Publishing.

Seager, Joni. 1993. *Earth Follies: Coming to Feminist Terms with the Global Environmental Crisis.* New York: Routledge.

References

Secor, D.H., V. Arefjev, A. Nikolaev and A. Sharov. 2000. 'Restoration of Sturgeons: Lessons from the Caspian Sea Sturgeon Ranching Programme'. *Fish and Fisheries* 1: 215–30.

Sen, Jai. 2018. *The Movements of Movements – Part 1: What Makes Us Move?* Oakland: PM Press.

Severson, Kim. 2016. 'A Rush of Americans, Seeking Gold in Cuban Soil'. *New York Times*, 20 June 2016. www.nytimes.com/2016/06/22/dining/cuba-us-organic-farming.html (accessed 9 January 2020).

Shaddick, G., M.L. Thomas, P. Mudu, G. Ruggeri and S. Gumy. 2020. 'Half the World's Population Are Exposed to Increasing Air Pollution'. *Nature Partner Journals Climate and Atmospheric Science* 3: 23. https://doi.org/10.1038/s41612-020-0124-2.

Shahgedanova, Maria, and Timothy P. Burt. 1994. 'New Data on Air Pollution in the former Soviet Union'. *Global Environmental Change* 4 (3): 201–27.

Shandra, John M., Michael Restivo and Jamie M. Sommer. 2019. 'Do China's Environmental Gains at Home Fuel Forest Loss Abroad? A Cross-national Analysis'. *Journal of World-Systems Research* 25(1), 83–110.

Shapiro, Judith. 2001. *Mao's War against Nature: Politics and the Environment in Revolutionary China*. Cambridge: Cambridge University Press.

Sharma, Ranju, Ngangbam Sarat Singh, Neha Dhingra and Talat Parween. 2020. 'Bioremediation of Oil-Spills from Shore Line Environment', in Mohammad Oves, Mohammad Omaish Ansari, Mohammad Zain Khan, Mohammad Shahadat and Iqbal M.I. Ismail (eds), *Modern Age Waste Water Problems*, pp. 275–91. Cham: Springer.

Shawki, Ahmed. 2006. *Black Liberation and Socialism*. Chicago: Haymarket Books.

Sheehan, Helena. 1985. *Marxism and the Philosophy of Science: A Critical History*. Atlantic Highlands: Humanities Press.

Shimp, Kaleb. 2009. 'The Validity of Karl Marx's Theory of Historical Materialism'. *Major Themes in Economics* 11 (11): 35–56.

Shixiong Cao, Li Chen, David Shankman, Chunmei Wang, Xiongbin Wang and Hong Zhang. 2011. 'Excessive Reliance on Afforestation in China's Arid and Semi-arid Regions: Lessons in Ecological Restoration'. *Earth-Science Reviews* 104 (4): 240–5.

Sidorchuk, Aleksey Y., and Valentin N. Golosov. 2003. 'Erosion and Sedimentation on the Russian Plain, II: The History of Erosion and Sedimentation during the Period of Intensive Agriculture'. *Hydrological Processes* 17: 3347–58.

Smil, Václav. 2015. *China's Environmental Crisis: An Inquiry into the Limits of National Development*. London: Routledge.

Smirnov, V.P. 2009. 'Tokamak Foundation in USSR/Russia 1950–1990'. *Nuclear Fusion* 50 (1): 014003.

Sojinu, Samuel O., Jizhong Wang, Oluwadayo Sonibare and Eddy Zeng. 2010. 'Polycyclic Aromatic Hydrocarbons in Sediments and Soils from Oil Exploration Areas of the Niger Delta, Nigeria'. *Journal of Hazardous Materials* 174 (1–3): 641–7.

Sokolovsky, Valentin. 2004. 'Fruits of a Cold War', in Johan Sliggers and Willem Kakebeeke (eds), *Clearing the Air: 25 years of the Convention on Long-Range Transboundary Air Pollution*, pp. 7–15. Geneva: United Nations.

Sosa, Dayana, Isabel Hilber, Roberto Faure, Nora Bartolomé, Osvaldo Fonseca, Armin Keller, Thomas D. Bucheli and Arturo Escobar. 2019. 'Polycyclic Aromatic Hydrocarbons and Polychlorinated Biphenyls in Urban and Semi-urban Soils of Havana, Cuba'. *Journal of Soils & Sediments: Protection, Risk Assessment, & Remediation* 19 (3): 1328–41.

Suing, Guillaume, and Viktor Dedaj. 2018. *L'Écologie Réelle: Une Histoire Soviétique et Cubaine*. [Real Ecology: A Soviet and Cuban History.] Paris: Delga.

Szalai, Erzsébet. 2005. *Socialism: An Analysis of Its Past and Future*. Budapest: Central European University Press.

Szymanski, Albert. 1979. *Is the Red Flag Flying? The Political Economy of the Soviet Union in the World Today*. London: Zed Books.

Taff, Gregory, Daniel Müller, Tobias Kuemmerle, Esra Ozdeneral and Stephen Walsh. 2009. 'Reforestation in Central and Eastern Europe after the Breakdown of Socialism', in Harini Nagendra and Jane Southworth (eds), *Reforesting Landscapes*, pp. 1–30. Dordrecht: Springer.

Tal, Alon. 2002. *Pollution in a Promised Land*. Berkeley: University of California Press.

Tao, Hong, Nannan Yu and Zhonggen Mao. 2019. 'Does Environment Centralization Prevent Local Governments from Racing to the Bottom? Evidence from China'. *Journal of Cleaner Production* 231: 649–59.

Taylor, Peter J. 1987. 'The Poverty of International Comparisons: Some Methodological Lessons from World-Systems Analysis'. *Studies in Comparative International Development* 22 (1): 12–39.

Tecle, Aregai. 2017. 'Downstream Effects of Damming the Colorado River'. *International Journal of Lakes and Rivers* 10 (1): 7–33.

Temper, Leah, Daniela del Bene and Joan Martinez-Alier. 2015. 'Mapping the Frontiers and Front Lines of Global Environmental Justice: The Ejatlas'. *Journal of Political Ecology* 22: 255–78.

Thomas, Valerie M., and Anna O. Orlova. 2001. 'Soviet and Post-Soviet Environmental Management: Lessons from a Case Study on Lead Pollution'. *Ambio* 30 (2): 104–11.

Trunov, Alexander. 2017. 'Deforestation in Russia and Its Contribution to the Anthropogenic Emission of Carbon Dioxide in 1990–2013'. *Russian Meteorology and Hydrology* 42: 529–37.

Tucker, Michael. 1995. 'Carbon Dioxide Emissions and Global GDP'. *Ecological Economics* 15: 215–23.

Tucker, Richard P. 2000. *Insatiable Appetite. The United States and the Ecological Degradation of the Tropical World*. Berkeley: University of California Press.

Turner Terisa E., and Leigh Brownhill. 2006. 'Ecofeminism as Gendered, Ethnicized Class Struggle: A Rejoinder to Stuart Rosewarne'. *Capitalism Nature Socialism* 17 (4): 87–95.

References

Turnock, David. 2006. *The Economy of East Central Europe, 1815–1989: Stages of Transformation in a Peripheral Region.* London: Routledge.

Tyner, James A. 2017. *From Rice Fields to Killing Fields: Nature, Life, and Labor under the Khmer Rouge.* Syracuse: Syracuse University Press.

UN Climate Change. 2020. 'Global Warming Potentials (IPCC Second Assessment Report)'. https://unfccc.int/process/transparency-and-reporting/greenhouse-gas-data/greenhouse-gas-data-unfccc/global-warming-potentials (accessed 22 October 2020).

UNSD (United Nations Statistics Division). 2007. 'Total Surface Area as of 19 January 2007'. https://unstats.un.org/unsd/environment/totalarea.htm (accessed 7 November 2020).

US EPA (2020a) 'Superfund: National Priorities List (NPL)'. www.epa.gov/superfund/superfund-national-priorities-list-npl (accessed 2 November 2020).

US EPA (2020b). 'Report on the Environment: Contaminated Land – What Are the Trends in Contaminated Land and Their Effects on Human Health and the Environment?' www.epa.gov/report-environment/contaminated-land (accessed 20 November 2020).

Van der A, Ronald J., Bas Mijling, Jieying Ding, Maria Elissavet Koukouli, Fei Liu, Qing Li, Huiqin Mao and Nicolas Theys. 2017. 'Cleaning up the Air: Effectiveness of Air Quality Policy for SO$_2$ and NO$_x$ Emissions in China'. *Atmospheric Chemistry and Physics* 17: 1775–89.

Vine, David. 2015. *Base Nation: How U.S. Military Bases Abroad Harm America and the World.* New York: Metropolitan Books.

Visser, Oane, and Max Spoor. 2011. 'Land Grabbing in Post-Soviet Eurasia: The World's Largest Agricultural Land Reserves at Stake'. *The Journal of Peasant Studies* 38 (2): 299–323.

Vltchek, Andre. 2015. *Exposing Lies of the Empire.* Jakarta: Badak Merah Semesta.

Vorobeichik, Evgenii, and S. Yu. Kaigorodova. 2017. 'Long-Term Dynamics of Heavy Metals in the Upper Horizons of Soils in the Region of a Copper Smelter Impacts during the Period of Reduced Emission'. *Eurasian Soil Science* 50: 977–90.

Wallerstein, Immanuel. 1979. *The Capitalist World Economy: Essays by Immanuel Wallerstein.* Cambridge: Cambridge University Press.

Wang Xuanxuan, Yaning Chen, Li Zhi, Gonghuan Fang, Wang Fei and Liu Haijun. 2020. 'The Impact of Climate Change and Human Activities on the Aral Sea Basin over the Past 50 Years'. *Atmospheric Research* 245: 105125.

Watts, Michael. 1997. 'Black Gold, White Heat', in Steve Pile and Michael Keith (eds), *Geographies of Resistance*, pp. 33–67. London: Routledge.

Weaver, Jace. 2014. *The Red Atlantic: America Indigenes and the Making of the Modern World, 1000–1927.* Chapel Hill: The University of North Carolina Press.

Weiner, Douglas R. 1988. *Models of Nature: Ecology, Conservation, and Cultural Revolution in Soviet Russia.* Bloomington and Indianapolis: Indiana University Press.

Weiner, Douglas R. 1999. *A Little Corner of Freedom: Russian Nature Protection from Stalin to Gorbachëv.* Berkeley: University of California Press.

Weiner, Douglas R. 2006. 'Environmental Activism in the Soviet Context: A Social Analysis', in Christof Mauch, Nathan Stoltzfus and Douglas R. Weiner (eds), *Shades of Green: Environmental Activism around the Globe*, pp. 101–33. Oxford: Rowman & Littlefield.

Weiner, Douglas R. 2009. 'The Predatory Tribute-Taking State: A Framework for Understanding Russian Environmental History', in Kenneth Pomeranz and Edmund Burke (eds), *The Environment and World History*, pp. 276–315. Berkeley; University of California Press.

Weiner, Douglas R. 2017. 'Communism and Environment', in Juliane Fürst, Silvio Pons and Mark Selden (eds), *The Cambridge History of Communism*, pp. 502–28. Cambridge: Cambridge University Press.

Weiner, Douglas, and John Brooke. 2018. 'Conclusions: Nature, Empire, and Intelligentsia', in Nicholas Breyfogle (ed.), *Eurasian Environments: Nature and Ecology in Eurasian History*, pp. 298–315. Pittsburgh: University of Pittsburgh Press.

Welford, Richard. 1991. 'The Environmental Impact of German Reunification'. *European Environment* 1: 8–11.

Welsh, Ian, and Andrew Tickle. 1998. 'The 1989 Revolutions and Environmental Politics in Central and Eastern Europe', in Andrew Tickle and Ian Welsh (eds), *Environment and Society in Eastern Europe*, pp. 1–29. Harlow: Longman.

Westing, Arthur H. 1984. *Herbicides in War: The Long-Term Ecological and Human Consequences*. London: Taylor and Francis.

White, Kristopher. 2013. 'Nature–Society Linkages in the Aral Sea Region'. *Journal of Eurasian Studies* 4: 18–33.

Whittle, Daniel, and Orlando Rey Santos. 2006. 'Protecting Cuba's Environment: Efforts to Design and Implement Effective Environmental Laws and Policies in Cuba'. *Cuban Studies* 37: 73–103.

WHO (World Health Organisation). 2019. 'Drinking-Water'. www.who.int/news-room/fact-sheets/detail/drinking-water (accessed 22 December 2020).

Wiedmann, Thomas, and Manfred Lenzen. 2018. 'Environmental and Social Footprints of International Trade.' *Nature Geoscience* 11: 314–21.

Wiedeman, Thomas, Manfred Lenzen, Lorenz T. Keyßer and Julia K. Steinberger. 2020. 'Scientists' Warning on Affluence'. *Nature Communications* 11: 3107.

Willer, Helga, Minou Yussefi-Menzler and Neil Sorensen. 2008. *The World of Organic Agriculture: Statistics and Emerging Trends 2008*. https://orgprints.org/13123/2/willer-yussefi-sorensen-2008-final-tables.pdf (accessed 9 January 2021).

Williams, Michael. 2006. *Deforesting the Earth: From Prehistory to Global Crisis – An Abridgment*. Chicago: The University of Chicago Press.

Wolfson, Ze'ev. 1994. *The Geography of Survival: Ecology in the Post-Soviet Era*. London: M.E. Sharpe.

Woodhouse, Connie. 2003. 'Droughts of the Past: Implications for the Future?' in Sherry L. Smith (ed.), *The Future of the Southern Plains*, pp. 95–113. Norman: University of Oklahoma Press.

References

World Bank. 2019. 'World Development Indicators', last updated 10 July 2019. http://data.worldbank.org/indicator/ EN.ATM.CO2E.PC (accessed 4 November 2020).

World Bank. 2020a. 'Trade (% GDP)'. https://data.worldbank.org/indicator/NE.TRD.GNFS.ZS (accessed 21 October 2020).

World Bank. 2020b. 'Arable Land (% Arable Land) – Cuba'. https://data.worldbank.org/indicator/AG.LND.ARBL.ZS?locations=CU (accessed 26 November 2020).

World Bank. 2021a. 'Forest Area (Sq. Km) – Burkina Faso'. https://data.worldbank.org/indicator/AG.LND.FRST.K2?locations=BF (accessed 13 January 2021).

World Bank. 2021b. 'Population, Total'. https://data.worldbank.org/indicator/SP.POP.TOTL?end=2019&start=1960 (accessed 13 January 2021).

Wu, Zijin. 1987. 'The Origins of Environmental Management in China', in Bernhard Glaeser (ed.), *Learning from China? Development and Environment in Third World Countries*, pp. 111–19. London: Allen & Unwin.

Wuepper, David, Pasquale Borrelli, Daniel Mueller and Robert Finger. 2020. 'Quantifying the Soil Erosion Legacy of the Soviet Union'. *Agricultural Systems* 185: 102940.

WWF (World Wildlife Federation). 2020. *Living Planet Report 2020: Bending the Curve of Biodiversity Loss*. Gland: WWF.

Wynn, Graeme. 2007. *Canada and Arctic North America: An Environmental History*. Santa Barbara: ABC-CLIO.

Xu, Zhun. 2018. *From Commune to Capitalism: How China's Peasants Lost Collective Farming and Gained Urban Poverty*. New York: Monthly Review Press.

Yaffe, Helen. 2019. *We Are Cuba! How a Revolutionary People Have Survived in a Post-Soviet World*. New Haven: Yale University Press.

Yanitsky, Oleg Nikolaevich. 2012. 'From Nature Protection to Politics: The Russian Environmental Movement 1960–2010'. *Environmental Politics* 21 (6): 922–40.

Ying, Irene, and Mike Carlowicz. 2016. 'Sulfur Dioxide Emissions Fall in China, Rise in India'. https://earthobservatory.nasa.gov/images/91270/sulfur-dioxide-emissions-fall-in-china-rise-in-india (accessed 20 November 2020).

Young, Phyllis. 1997. 'Beyond the Water Line', in Jace Weaver (ed.), *Defending Mother Earth: Native American Perspectives on Environmental Justice*, pp. 85–98. Maryknoll: Orbis Books.

Zecchini, Salvatore. 1996. *Environmental Information Systems in the Russian Federation: An OECD Assessment*. Paris: OECD.

Zhang, Ke, John A. Dearing, Terence P. Dawson, Xuhui Dong, Xiangdong Yang and Weiguo Zhang. 2015. 'Poverty Alleviation Strategies in Eastern China Lead to Critical Ecological Dynamics'. *Science of the Total Environment* 506–7: 164–81.

Zhang, Qiang, Xujia Jiang, Dan Tong, Steven J. Davis, Hongyan Zhao, Guannan Geng, Tong Feng, Bo Zheng, Zifeng Lu, David G. Streets, Ruijing Ni, Michael Brauer, Aaron van Donkelaar, Randall V. Martin, Hong Huo, Zhu Liu, Da Pan, Haidong Kan, Yingying Yan, Jintai Lin, Kebin He and Dabo Guan. 2017. 'Transboundary Health Impacts of Transported Global Air Pollution and International Trade'. *Nature* 543: 705–9.

Zhou, Yang, and Yansui Liu. 2018. 'China's Fight against Soil Pollution'. *Science* 362: 298.

Zhou, Yu, Li Zhang, Rasmus Fensholt, Kun Wang, Irina Vitkovskaya and Feng Tian. 2015. 'Climate Contributions to Vegetation Variations in Central Asian Drylands: Pre- and Post-USSR Collapse'. *Remote Sensing* 7 (3): 2449–70.

Ziegler, Charles E. 1980. 'Soviet Environmental Policy and Soviet Central Planning: A Reply to McIntyre and Thornton'. *Soviet Studies* 32 (1): 124–34.

Zierler, David. å2011. *The Invention of Ecocide: Agent Orange, Vietnam, and the Scientists Who Changed the Way We Think about the Environment*. Athens: University of Georgia Press.

Zonn, Igor, Michael H. Glantz and Alvin Rubinstein. 1994. 'The Virgin Lands Scheme in the Former Soviet Union', in Michael H. Glantz (ed.), *Drought Follows the Plow: Cultivating Marginal Areas*, pp. 135–50. Cambridge: Cambridge University Press.

Index